AF905548

# Ab-initio Assisted Petrology

Giulio Armando Ottonello · Marco Attene ·
Marino Vetuschi Zuccolini

# Ab-initio Assisted Petrology

Springer

Giulio Armando Ottonello
Fellow of the Mineralogical Society
of America
Corresponding Member of the Accademia
Nazionale dei Lincei
Rome, Italy

Marco Attene
CNR
Istituto di Matematica Applicata e
Tecnologie Informatiche
Genova, Italy

Marino Vetuschi Zuccolini
DISTAV
Università di Genova
Genova, Italy

ISBN 978-3-032-25657-7 ISBN 978-3-032-25658-4 (eBook)
https://doi.org/10.1007/978-3-032-25658-4

© The Editor(s) (if applicable) and The Author(s), under exclusive license to Springer Nature Switzerland AG 2026

This work is subject to copyright. All rights are solely and exclusively licensed by the Publisher, whether the whole or part of the material is concerned, specifically the rights of translation, reprinting, reuse of illustrations, recitation, broadcasting, reproduction on microfilms or in any other physical way, and transmission or information storage and retrieval, electronic adaptation, computer software, or by similar or dissimilar methodology now known or hereafter developed.
The use of general descriptive names, registered names, trademarks, service marks, etc. in this publication does not imply, even in the absence of a specific statement, that such names are exempt from the relevant protective laws and regulations and therefore free for general use.
The publisher, the authors and the editors are safe to assume that the advice and information in this book are believed to be true and accurate at the date of publication. Neither the publisher nor the authors or the editors give a warranty, expressed or implied, with respect to the material contained herein or for any errors or omissions that may have been made. The publisher remains neutral with regard to jurisdictional claims in published maps and institutional affiliations.

This Springer imprint is published by the registered company Springer Nature Switzerland AG
The registered company address is: Gewerbestrasse 11, 6330 Cham, Switzerland

If disposing of this product, please recycle the paper.

# Preface

An enormous amount of experimental and theoretical work has been performed in petrology over the last few decades. Regrettably, however, one cannot claim we have arrived yet at a complete understanding of the way in which igneous rocks form as a result of partial melting of their deep source and then of fractional crystallization of magma. The main reason is that the conditions of temperature, pressure, and chemical composition relevant to natural processes are far too wide to be readily accounted for. This is especially the case for high-pressure conditions, at which phase equilibria experiments are demanding and calorimetric measurements virtually impossible. To give a single example, the current literature is controversial even for such a simple substance as MgO (periclase), for which discrepancies of hundreds of degrees are found between the high-pressure melting temperatures determined in various ways.

Under such circumstances, first-principles theoretical calculations represent an attractive alternative as has already been extensively documented (e.g., Ottonello, 2024). The present publication is thus another plea to foster the application of Quantum Mechanics to petrology in the special case of $CaO$-$MgO$-$Al_2O_3$-$SiO_2$ system, which owes its fundamental importance to the fact that it is the simplest system of direct relevance to natural igneous processes. Such approaches have become feasible because silicate melts may be treated as simple dielectrics whose static and vibrational properties may be accurately depicted through ab-initio procedures (Ottonello and Richet, 2014).

Technically, as will be explained in detail, the method followed is to couple the computation procedures of the Density Functional Theory (DFT) for representative clusters with hybrid functionals with those of the Polarized Continuum Model and Scaling Particle Theory (Tomasi and Persico, 1994; Reiss, 1965; Pierotti, 1976). In addition, the need to deal with high-P conditions forces one to have a complete appraisal of vibrational anharmonicity, which can be achieved only through quantum-mechanical procedures. These computations are quite demanding, but the benefits far exceed the efforts. For instance, one can retrieve from vibrational calculations the isobaric heat capacity of a substance ($C_P$), and, conversely, investigate whether or not a given $C_P = f(T)$ functional form violates the Dulong-Petit constraint with dramatic effects in the computation of the Gibbs free energy (G) and its decoupling in enthalpic (H) and entropic (S) counterparts. Among the extensive data already gathered by Earth scientists, special mention must be made of the high-quality calorimetric data produced and of the vast efforts carried out in the buildup of accurate phase diagrams. These *Schreinemaker's projections* of equipotential loci to depict the extrinsic stability limits of the various nucleating substances have offered the opportunity of empirical parameterization of mixing rules and the birth of models based on inverse approaches. The mathematical flexibility of the various models that appeared through the years allows, in many circumstances, an almost perfect reproducibility of the observables, in terms of composition, T, P, and aggregations. Unfortunately, however, many thermodynamic parameters currently adopted for the various nucleating phases cannot survive a strict first-principles analysis.

A last point to be considered is the mathematical approach adopted to take into account chemical complexity. Here we deal with a third-order simplex, i.e., a system of four compositional units assumed as sufficient and appropriate to form all aggregation states (oxides). The construction begins with the assessment of the *melting curves* of the four pure oxides (i.e., the P, T loci of equilibrium between liquid and crystalline state, intrinsically affected by the polymorphic transitions), then the six first-order simplexes (i.e. *binaries* in the petrological jargon), then the four second-order simplexes (*thernaries*), to end with the single third-order *quaternary*. In this progressive construction, it is crucial to adopt an appropriate fragmentation of the compositional space (*Delaunay fragmentation*) and the convex-hull analysis of equilibrium loci. The arising fragments (Voronoi cells) bear analogies with the *Brillouin zone* (Brillouin, 1926) familiar to Crystallographers. Being in a multidimensional domain, the lifted *Delaunay triangulation* of the samples allows to realize a piecewise-linear interpolation and to search the *convex hull ch*(S) of the set of points S, i.e., the smallest convex set containing S.

Increasing the chemical complexity (i.e., the rank of the system). We lose the capability of transferring the information on a sheet of paper. Whereas for a first-rank simplex at constant P we may reproduce the phase stability conveniently limited with a x-y plot (x = composition and y = T), for a second-rank system at constant P we may adopt lifted coordinates (for an A-B-C system $x = A + 0.5B$; $y = 0.866C$), but T must be superimposed on the diagram as regularly spaced isotherms. For a third rank simplex at constant P, the information on T resides in the outputs generated by the computational package, and the form of the space (i.e., a tetrahedron generated by barycentric coordinates) may be seen in real time on the computer screen or reproduced through *screenshot* realizations. The first realizations of sorts of screenshots are due to Eitel (1951) and are an emblematic representation of the intuitive power of this Scientist.

The advantage of a modern numerical analysis with respect to these pioneering representations resides in the post-processing: applying the petrology rules, we may retrieve from the database the petrogenetic evolution in an open system of a given starting composition retrieving at each point of interest all thermodynamic and thermophysical parameters of the nucleating phases and the energy balances at trivariant, divariant, univariant and invariant points.

More poetically, we may assert that we are able to assign a precise *form* to a precise *energy*.

Calci (PI), Italy — Giulio Armando Ottonello
Genova, Italy — Marco Attene
Genova, Italy — Marino Vetuschi Zuccolini

**Acknowledgments** The manuscript benefited of the crytical reviews and suggestions of Jibamitra Ganguly and Pascal Richet which are gratefully acknowledged.

**Competing Interests** The authors have no competing interests to declare that are relevant to the content of this manuscript.

# Contents

# Chapter 1
# Premises

Building a multicomponent phase diagram is a rigorous stepwise procedure that begins with the assessment of the thermodynamic-thermophysical properties of the pure components under all polymorphic and aggregation states. The parameters, initially based on experiments, must prove to obey first principles, to be internally consistent, and to be valid in a sufficiently wide thermobaric range. In petrological applications, it is particularly stringent, the assessment of the "*melting curve*" of the substance, depicting the effect of isotropic pressure (P) on the melting temperature (T). The construction then proceeds with the investigation of all binary interactions, then the ternary ones, the quaternary, etc., up to the rank of the system. The compositional space is finally discretized into appropriate regions in order to explore Schreiner's projections of equipotential loci conveniently.

## 1.1 Introduction

The CaO-MgO-$Al_2O_3$-$SiO_2$ system (hereafter CMAS), when investigated at isothermal and isobaric conditions, is a third-rank simplex whose representation requires the adoption of simplicial-mesh techniques coupled with Delaunay fragmentation and convex-hull analysis of equilibrium loci. This analysis, in its turn, requires a quite stringent thermodynamic investigation of all phases eventually nucleating in the system at the various P, T conditions of interest. The usual visualization of the phase topology of CMAS is obtained through tetrahedral coordinates, which are a convenient translation of barycentric coordinates. Seminal papers on the way of treating third-rank simplexes where provided by Boecke (1916), Vogel (1937), Hanemann and Schrader (1943) but these, though of excellent quality, were limited to discussions of sections parallel to one side of the tetrahedron, The first compendium on quaternary systems is due to Wilhelm Eitel, former Director of the Kaiser-Wilhelm Institute for silicate research. His book, first published in Leipzig in

© The Author(s), under exclusive license to Springer Nature Switzerland AG 2026

G. A. Ottonello et al., *Ab-initio Assisted Petrology*,
https://doi.org/10.1007/978-3-032-25658-4_1

1943, was totally destroyed during the second-world war, with the exception of one copy. A translation in English, due to J. G. Phillips and T. G. Madgwick, appeared in 1951. Eitel (1951) is, as far as we know, the first available comprehensive treatment of quaternary systems, offering important clues on the quantitative assessment of melting and/or crystallization relations.

A quantitative analysis of third-rank simplexes must begin with a detailed outline of the melting behavior of the pure components at various pressures (for geological investigations, from $10^5$ Pa to P regimes typical of planetary interior conditions, i.e., up to hundreds of GPa). One must proceed then with the six binary joins limiting the tetrahedron, then the four ternary faces, to end up with the four-component volume. In the case of CMAS, more than one-hundred phases may nucleate in the system, depending on the current P, T conditions, and some of them are mixtures of two or more components. The melt itself may be seen as a mixture, which, in some circumstances (i.e., low P and high acidity), is prone to decompose. Obviously, with such topological complexity, it is almost compulsory to avoid, at best, all sorts of empiricism and to rely, whenever possible, on first principles. To achieve sufficient precision, the compositional space must be spanned at regular intervals. The *lifted Delaunay triangulation* (see later) needs, in fact, limited compositional increments (i.e., at per-cent or per-mil level). T-steps of 10 K or less are appropriate. Pressure must be imposed as constant, in order to represent Schreinemaker's isobaric projections of phase boundaries. The *Voronoi cells* of the simplicial mesh may then be treated by convex-hull analysis to depict the loci of trivariant, divariant, univariant, and invariant equilibrium. Repeated smoothing steps must be performed to reduce the numerical noise. Finally, the plethora of self-generated energy data files must be explored by post-processing codes able to retrieve the lines of descent at closed-system conditions and to return salient thermodynamic information (i.e., latent heats, entropies, molar volumes, buoyancy, etc.).

# Chapter 2
# Simplicial Complexes, Convex Hulls, and Delaunay Triangulations

A convenient way to define a specific composition is to define the relative amount of each component. The set of all the possible compositions of k pure components can be modeled as an abstract (k−1)-simplex: in such a simplex, each vertex represents a pure component, and each specific composition is defined by k numbers, each specifying the amount of one of the pure components. When the simplex has a geometric realization (e.g., when it is visualized), a specific composition corresponds to a unique point of the Euclidean simplex that can be specified through a linear combination of the vertices. When a function on a multidimensional domain (e.g., a liquidus surface) is sampled at a finite set of points S, the lifted *Delaunay triangulation* of the samples is realized through a piecewise-linear interpolation. The *convex hull ch*(S) of a set of points S is the smallest convex set containing S.

## 2.1 Abstract Simplicial Complex

Let V be a finite nonempty set. An *abstract simplicial complex* (Attene 2004) is a collection Σ of finite nonempty subsets of V such that every $v \in V$ belongs to Σ and if A is an element of Σ, then so is every nonempty subset of A. Each element of V is called a *vertex* of Σ. An element σ of Σ of cardinality k + 1 is called a *k-simplex*, and k is the *order* of σ (hence, a 0-simplex is a vertex). A d-simplex σ is said to be *incident* at a k-simplex τ if $\tau \subseteq \sigma$ and, in this case, τ is said to be a *face* of σ.

Let us assume to be in the Euclidean space $\mathbf{R}^n$, i.e., $R^n$ with the Euclidean metric. An Euclidean simplex of order $k \leq n$ (*k*-simplex) is the locus of the points in $\mathbf{R}^n$ that can be expressed as the convex combination of $k + 1$ points $\{p_1, p_{k+1}\}$ in general position (arbitrarily expanded from 298.15 to 5000 K). Hence, an Euclidean *k*-simplex is the union of all the points p *P* that can be expressed as:

© The Author(s), under exclusive license to Springer Nature Switzerland AG 2026 
G. A. Ottonello et al., *Ab-initio Assisted Petrology*,
https://doi.org/10.1007/978-3-032-25658-4_2

$$P = \sum_{i=1}^{k+1} l_i p_i \tag{2.1}$$

$$\sum_{i=1}^{k+1} l_i = 1, \quad l_i \in \Re, \quad l_i \geq 0 \tag{2.2}$$

The $k + 1$ points $p_1, \ldots, p_{k+1}$ are in general position (or affinely independent) if the $k$ vectors $\mathbf{p}_2 - \mathbf{p}_1, \mathbf{p}_3 - \mathbf{p}_1, \ldots, \mathbf{p}_{k+1} - \mathbf{p}_1$ are linearly independent.

Any Euclidean $d$-simplex $\tau$ of cardinality $d < k$ is called a *d-face* (or simply a *face*) of an Euclidean $k$-simplex $\sigma$ if $\tau \subset \sigma$. A finite collection $\Sigma$ of Euclidean simplexes is an Euclidean simplicial complex if and only if the following conditions hold:

i. for each simplex $\sigma$ of $\Sigma$, all faces of $\sigma$ belong to $\Sigma$;
ii. for each pair of simplexes $\sigma$ and $\tau$, either their intersection is empty, or it is a face of both $\sigma$ and $\tau$.

If $k$ is the maximum of the orders of the simplexes of $\Sigma$, then k is the order of $\Sigma$, and $\Sigma$ is an Euclidean $k$-simplicial complex.

Let now V be a finite nonempty set and let $\Sigma$ be an abstract simplicial complex on V. An *embedding* of $\Sigma$ into $\mathbf{R}^n$ is a function $f$: V $\rightarrow \mathbf{R}^n$, such that:

(1) For every $k$-simplex $\sigma$ in $\Sigma$, the set $\sigma$' generated by the images of the vertices of $\sigma$ through $f$ is an Euclidean $k$-simplex;
(2) The collection $\Sigma$' of all the simplices generated by $f$, as described above, fulfills condition (ii) of the definition of Euclidean simplicial complex.

Thus, an embedding of an abstract simplicial complex is fully defined by mapping its vertices into points in Euclidean space. Furthermore, requirement (1) states that the set of vertices of each simplex must be projected into a set of linearly independent points. Thus, the dimension $n$ of the space where the complex is embedded must be at least as large as the order of the complex (Attene 2004).

If $f$ is a bijection and $S \subseteq \mathbf{R}^3$ is the union of all the Euclidean $k$-simplexes generated by the vertices of $f(\sigma)$, for each $\sigma \in \Sigma$, then the set S is a *geometric realization* of $\Sigma$, and it is denoted as $|\Sigma|$.

An Euclidean simplicial complex ($k$-simplicial complex) is also said to be a *simplicial mesh* ($k$-dimensional simplicial mesh). Two- and three-dimensional simplicial meshes are also called *triangle* and *tetrahedral* meshes, respectively.

## 2.2 Barycentric Coordinates

A convenient way to define a specific composition is to define the relative amount of each component. The set of all the possible compositions of $k$ pure components can be modeled as an abstract $(k-1)$-simplex: in such a simplex, each vertex represents a pure component, and each specific composition is defined by $k$ numbers, each specifying the amount of one of the pure components. When the simplex has a geometric realization (e.g., when it is visualized), a specific composition corresponds to a unique point of the Euclidean simplex that can be specified through a linear combination of the vertices. *Barycentric coordinates* identify points in a $k$-simplex independently of its embedding. When expressed in barycentric coordinates, each point is identified by k + 1 real numbers ($\alpha_0$, ..., $\alpha_k$) such that $\alpha_0 + \alpha_1 + \ldots + \alpha_k = 1$ and $\alpha_i \geq 0 \quad \forall i = 0, \ldots, k$.

When a $k$-simplex has a geometric realization given by the position of its vertices ($\mathbf{v}_0, \ldots, \mathbf{v}_k$) in $\mathbf{R}^n$ $n \geq k$, the geometric realization of the barycentric point ($\alpha_0$, ..., $\alpha_k$) is the convex combination.

$$\mathbf{p} = \alpha_0 \mathbf{v}_0 + \alpha_1 \mathbf{v}_1 + \ldots + \alpha_k \mathbf{v}_k \tag{2.3}$$

To compute the barycentric coordinates of an Euclidean point lying in an Euclidean simplex, it is possible to *subdivide* the simplex at that point, and to evaluate the amount of area (or volume, or hyper volume, ...) of each of the subdivisions with respect to the total area (or volume, or hyper volume, ...) of the simplex (see Fig. 2.1).

For example, we may consider the bi-dimensional case of a point p in a triangle. The barycentric coordinates of p are the ratios between areas detected from p and the total area of the triangle. Thus, with reference to 1, p is represented in barycentric coordinates as:

$$(\alpha_0, \alpha_1, \alpha_2) = \left( \frac{A_0}{A}, \frac{A_1}{A}, \frac{A_2}{A} \right) \tag{2.4}$$

where $A_0, A_1, A_2$ are the areas of the triangle subdivisions and $A$ is the whole area of the triangle.

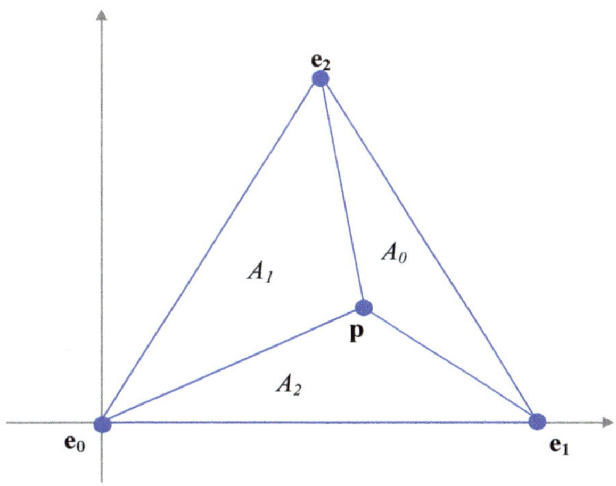


**Fig. 2.1** Geometric realization for the point p in the two-dimensional case

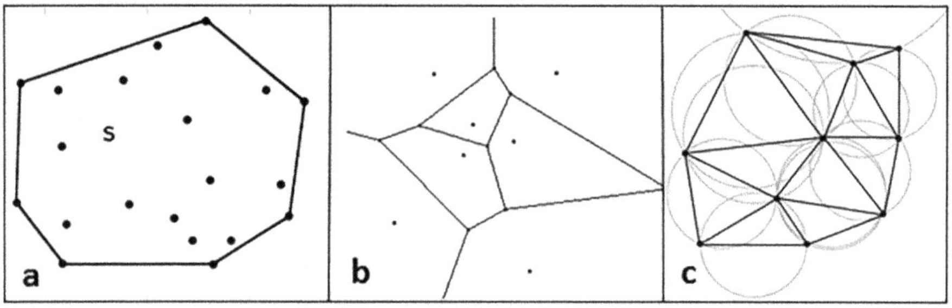


**Fig. 2.2** Convex hull (**a**), Voronoi diagram (**b**), and Delaunay triangulation (**c**). From Natali et al. (2011) with modifications

The above example can be generalized to higher dimensions by considering volumes or hyper-volumes instead of areas.

## 2.3 Convex Hulls and Delaunay Triangulations

When a function of one variable is sampled at a finite set of points, a convenient practice to approximate the function starting from the samples (if no other information is known) is to employ a linear interpolation. Analogously, when a function on a multidimensional domain (e.g., a liquidus surface) is sampled at a finite set of points, it can be useful to compute the lifted *Delaunay triangulation* of the samples to realize a piecewise-linear interpolation.

Let S be a set of points in Euclidean space with no accumulation points. The set of all points closer to a point **s** (*site*) of S than to any other point of S is the interior of a convex polytype called the *Dirichlet domain* or *Voronoi cell* for **s**. The set **Vor**(S) of Voronoi cells tessellates the whole space, and is named the **Voronoi** diagram of the set S (an example is in Fig. 2.2b).

The *convex hull ch*(S) of a set of points S is the smallest convex set containing S (Preparata and Shamos, 1985).

Herewith, we call a *facet* each primitive element constituting a convex hull. Thus, in *2d* a facet denotes a segment, in *3d* a flat polygon, etc....

Formally, the Delaunay triangulation is defined as the *dual graph* of the Voronoi diagram. In 2D, all the finite regions of the Voronoi diagram can be obtained by connecting the centers of the circumcircles of adjacent Delaunay triangles. The Delaunay triangulation can be characterized in terms of one of its properties, sometimes called the *Delaunay condition*:

- A $d$-dimensional simplicial mesh in $\mathbf{R}^d$ is a Delaunay triangulation if the interior of the circumsphere of each of its $d$-simplexes does not contain any vertex (see Fig. 2.2c).

The problem of finding the Delaunay triangulation/Voronoi diagram of a set of points in $n$-dimensional Euclidean space can be converted to the problem of finding the convex hull of a set of points in $(n + 1)$-dimensional space. A way to do that is to assign to the $(n + 1$-)th coordinate of each point p the value $\|p\|^2$; then the bottom side of the convex hull [.e. the facets whose normal vector has a negative $(n + 1)$-th coordinate] can be simply projected back to $n$-dimensional space by deleting the last coordinate. Provided that the facets of the convex hull are simplices (if not, they can be triangulated), the projection is a Delaunay triangulation (Aurenhammer 1991). In general, the projection of the lower hull of a $(d + 1)$-dimensional $k$-set polytype is a $d$-dimensional Delaunay triangulation (Schmitt and Spehner 2006). This generalization to higher dimensions will be used in the remainder to efficiently and robustly derive phase diagrams of systems made of an arbitrary number of components.

# Chapter 3
# Liquidus Derivation and Analysis

Within our architecture, we represent the space of chemical potentials in a d-simplex, with $d + 1$ corresponding to the total count of oxides (for instance, $d + 1 = 4$ in the CMAS system). During the analysis, we calculate the Gibbs free energy across a grid of evenly distributed sample points that cover the entire simplex. After determining the Gibbs free energy for every phase at a constant pressure (P) and specific temperature intervals (T), we utilize the convex-hull approach. This allows us to identify the specific compositions that maintain equilibrium across various phase states at a particular temperature. Because these data points are gathered from discrete locations across the composition domain, we employ Delaunay triangulation to interpolate the results. This creates a piecewise-linear model that approximates the naturally continuous *liquidus* surface.

## 3.1 Overview

Crucial in our analysis is the thermodynamic and thermophysical information on the pure components that, under various aggregation states, conform all the materials that potentially, at the various P, T conditions of interest, generate physically distinguishable regions (phases). Generally speaking, we may divide the phases that may form in a system as pertaining to three distinct classes: (a) mechanical mixtures; (b) mixtures with partial miscibility; (c) mixtures with complete miscibility. The term "*mixture*" (IUPAC notation) stresses that we treat all components in a mixture in the same manner (i.e., same standard state of pure component in pure phase at T, P of interest). Mechanical mixtures do not actually mix in any proportion; partial mixtures have only a limited solubility, while mixtures with complete miscibility accept all components in any proportion. Most of the crystalline phases may be treated as mechanical mixtures, and the liquid as a phase with complete miscibility over most of the compositional range (though with a discrete P–T-dependent miscibility gap in

© The Author(s), under exclusive license to Springer Nature Switzerland AG 2026 
G. A. Ottonello et al., *Ab-initio Assisted Petrology*,
https://doi.org/10.1007/978-3-032-25658-4_3

the high-acidity region of the system; see later). For the sake of simplicity, all realizations may be strictly isobaric, and the effect of P may be established *a-posterior*, by comparative analysis.

## 3.2 Step 1: Modeling the Gibbs Energy for Each Phase

In mechanical mixtures, there are no Gibbs free energy of mixing contributions, hence their Gibbs free energy is simply defined in terms of their standard-state properties (Tables 4.1 and 4.2) and heat capacity integrals:

$$G_{f,T} = H^0_{f,Tref} + \int_{T_{ref}}^{T} C_P dT - T\left(S^0_{Tref} + \int_{Tref}^{T} \frac{C_P}{T} dT\right) \tag{3.1}$$

A single computation, based on Eq. (3.1) (when working at ambient pressure) and stemming from the standard-state thermodynamic data (Table 4.1), will be generally sufficient for each mechanical mixture nucleating in the system of interest. Solid mixtures deserve an accurate treatment of their ideal and excess mixing terms:

$$G_T = \sum_i \mu_i^0 X_i + G_{mix,ideal} + G_{mix,excess} \tag{3.2}$$

with $\mu_i^0$ as chemical potential of the pure i$^{th}$ component in the mixture. We give later on some clues on the adopted procedures.

Silicate liquid exhibits, however, complete miscibility at sufficiently high-T, with miscibility gaps limited to the high-$SiO_2$-content regions (see later). For the liquid, we will consider two distinct (but operationally similar approaches): the Modified Quasi-Chemical Model (hereon MQC) and the Hybrid Polymeric Approach (HPA). MQC is a well-tested model with important applications in Material Science, but it has the disadvantage of requiring high-order interactions whose significance cannot be foreseen by the application of first principles. HPA is perhaps conceptually more complex, but it has the advantage of providing a conceptual rationale in the interpretation of the various forms of energy which concur to the overall stability of the amorphous substance.

For computational purposes, it is convenient to select a sampling grid in the system spanning in a uniform way the simplex. The adopted compositional step will depend upon the geometrical complexity of the liquidus and on the degree of accuracy one wants to achieve in terms of thermal assessment of divariant, univariant, and invariant points. In the case study we present here (the $CaO$–$MgO$–$Al_2O_3$–$SiO_2$ system here on CMAS), we sample the simplex along equally spaced points differing at the percent molar level. The simple algorithm adopted at the molar percent level is the following:

```
NSTEPS=100
NPOINT=NSTEPX+1
XINCRE=1.0/NSTEPX
  DO 10 i=1,npoint
       DO 20 ii=0,npoint-i
              DO 30 iii=0,npoint-i-ii
              xSiO2(i)=(i-1)*xincre
              xAl2O3(i)=ii*xincre
              xMgO(i)=iii*xincre
              xCaO(i)=1.0-xSiO2(i)-xAl2O3(i)-xMgO(,i)
30 CONTINUE
20 CONTINUE
10 CONTINUE
```

This algorithm can be easily extended to higher dimensions by adding further nested loops. A typical explorative calculation carried out at P = $10^5$ Pa, in the 1000–3000 °C range with $\Delta T = |10|$ steps and 0.01 molar segmentation turns out to generate ~$3.5 \times 10^7$ energy points.

## 3.3 Step 2: Deriving the Liquidus Samples

Our method for constructing equilibrium loci—specifically, isobaric Schreinemaker's projections of equipotentials—is exceptionally stable. To our knowledge, existing implementations have not demonstrated functionality for systems exceeding three components. While ternary system applications of the convex-hull approach are documented in literature (e.g., Voskov and Voronin 2010) and available in software prototypes like Matforge, our pipeline utilizes a dimension-independent version of the algorithm.

The process begins by aggregating the Gibbs free energies for all phases at a specific T and P. These points are mapped into a common coordinate system where the final coordinate represents the chemical potential, and the preceding coordinates define the composition within the $n$-component simplex. Formally, for $n$ components, a point $x$ is defined by $n - 1$ Euclidean coordinates. Our dataset S consists of vectors $S_i = <e_1, e_2, \ldots, e_{n-1}, G_\gamma(<e_1, e_2, \ldots, e_{n-1}>)>$, where $G_\gamma$ is the Gibbs free energy of phase $\gamma$ at that composition.

We then compute the convex hull Ch(S) of this point cloud and triangulate any non-triangular facets. By selecting only facets with downward-pointing normals (i.e., where the last coordinate of the normal is negative), we extract the "lowerhull" Lh(S). Projecting this lower hull back onto the compositional domain by dropping the energy coordinate yields a Delaunay triangulation, D. In this triangulation, vertices are classified as either "liquid" or "solid" based on the phase they represent. Consequently, each triangle in D identifies a region that—at the given T and P—corresponds to a single-phase field or a multi-phase equilibrium.

To reconstruct the liquidus section for a specific T and P, we identify all edges in D that bridge a liquid point and a solid point. The liquid–solid interface is essentially traced by liquid vertices connected to at least one solid vertex. Identifying coexisting liquids is more complex because we use discrete sampling; an edge between two liquid points might represent a true miscibility gap or simply be an artifact of the sampling grid on a single energy surface. To distinguish real coexistences, we compare D against a standard Delaunay triangulation K of **all** sample points (including those internal to the hull). We only confirm liquid–liquid coexistence if an edge exists in the lower hull triangulation D that is absent in the total sampling triangulation K.

## 3.4 Step 3: Interpolating the Points to Form the Liquidus Surface

By iterating the procedures in step 2 across the full range of required temperatures and pressures, we can interpolate the collected data points to establish a clear topology of the phase diagram. We continue to utilize Delaunay triangulation for this purpose, as it ensures the creation of well-proportioned simplexes that facilitate reliable numerical analysis. To demonstrate this, we applied the interpolation method to liquidus samples calculated for the CAS system—which represents one of the four boundary faces of the CMAS tetrahedral composition space. This specific model covers a temperature spectrum from 1400 to 2600 °C at a constant pressure of $10^5$ Pa. The resulting surface is illustrated in Fig. 3.1.

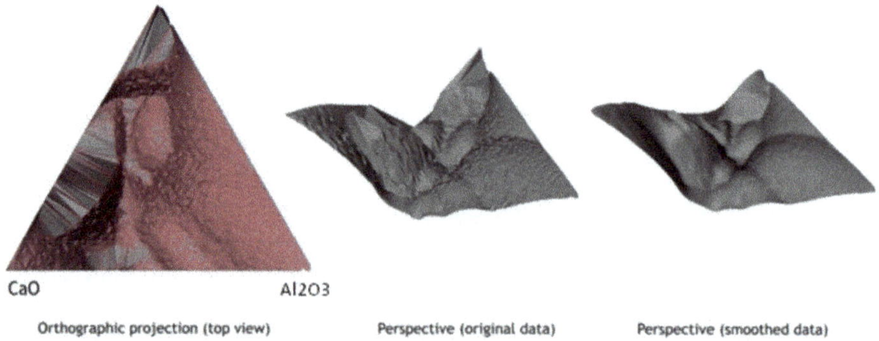


**Fig. 3.1** *Liquidus* of a CAS system in the range between 1400 and 2600 °C at a constant pressure of $10^5$ Pa. The smooth surface on the right was obtained by averaging the position of each vertex with the position of its neighbors in the mesh (so-called Laplacian smoothing)

# Chapter 4
# Thermodynamic and Thermophysical Assessment

As stressed from the very beginning, in our analysis, it is crucial that the thermodynamic and thermophysical information on the pure components and of the solid and liquid mixtures that, under the various aggregation states, conform all the materials that potentially nucleate, at the various P, T conditions of interest, generating physically distinguishable regions (phases). To attain this goal, we must rely on first principles and convey into formal statements (i.e., thermodynamic and thermophysical magnitudes) the complex structure-energy relations taking place with the modification of intensive variables.

## 4.1 The Parameters

The adopted isobaric heat capacity function is the Shomate equation:

$$C_P = a + b \times T + c \times T^{-2} + d \times T^{-1/2} + e \times T^{-3} + f \times T^2 + g \times T^3 + h \times T^{-1} \tag{4.1}$$

In the harmonic approximation, the isochoric heat capacity, $C_v$, of one mole of a crystalline solid is given by:

$$C_V = \frac{3R}{Z}\left(\frac{2}{\pi}\right)^3 \sum_{i=1}^{3} \int_0^X \frac{\left[\arcsin(X/X_i)^2\right] X^2 e^X dX}{\left(X_i^2 - X^2\right)^{1/2}\left(e^X - 1\right)^2} + \frac{R}{Z}\sum_{4}^{3n} e^{X_i}\left(\frac{X_i}{e^{-X_i} - 1}\right)^2 \tag{4.2}$$

The above equation disregards phonon dispersion; moreover, n is the number of atoms in the unit cell, $X_i$ is the ith a-dimensional frequency, and Z is the number of formula units per unit cell.

© The Author(s), under exclusive license to Springer Nature Switzerland AG 2026

G. A. Ottonello et al., *Ab-initio Assisted Petrology*,
https://doi.org/10.1007/978-3-032-25658-4_4

Following (Kieffer 1979a, b), the acoustic contributions at $\mathbf{k}_{max}$ (first term on the right of Eq. (8) are represented in terms of a sine-wave dispersion relation. The contribution of all remaining (optic) modes is grouped in the second term.

The isobaric heat capacity $C_P$ is readily obtained from the isochoric counterpart $C_V$ by using the standard relation:

$$C_P = C_V + T\alpha_T^2 K_{P,T} V_{P,T} \tag{4.3}$$

$\alpha_T$ in the above equation is the isobaric thermal expansion at T. Obviously, $C_V$ must obey the Dulong-Petit limit at high temperature.

The most convenient functional form for $\alpha_T$ is a polynomial in T, with exponents decreasing from 1 to $-3$ (Saxena et al. 1993):

$$\alpha_V = \alpha_0 T + \alpha_1 + \alpha_2 T^{-1} + \alpha_3 T^{-2} + \alpha_4 T^{-3} \tag{4.4}$$

Treating all vibrational normal modes as Einstein oscillators, the quasi-harmonic (QHA) gamma-mode analysis of the $\alpha_T K_T$ product gives:

$$\alpha_T K_T = -\left(\frac{\partial V}{\partial T}\right)_{P=0} \times \left(\frac{\partial V}{\partial P}\right)_{T=0} = \frac{R}{ZV}\sum_{i=4}^{3n} \gamma_i e^{-X_i}\left(\frac{X_i}{e^{X_i}-1}\right)^2 \tag{4.5}$$

where $\gamma_i$ is the Gruneisen gamma of the ith normal mode $\nu_I$; (i.e. $\gamma_i = \partial \ln \nu_i / \partial \ln V$).

The thermal dependence of the compressibility at zero pressure is here represented by a second-order dependence on T:

$$K_T = K_0 + K_1 T + K_2 T^2 \tag{4.6}$$

It is quite obvious that the bulk modulus $K_0$ implicit in Eq. (4.6) is that one ruling at the athermal limit. Often, however, in many cases, one may observe in literature the adoption of values obtained at room conditions, which is quite misleading and a source of errors and incomprehension.

A few additional parameters are added to account for the coupled effect of the intensive variables and to give mathematical elasticity to the procedure:

$$K' = \left(\frac{\partial K_{(0,T)}}{\partial P}\right)_{T=0} \tag{4.7}$$

$$K'' = \left(\frac{\partial K'}{\partial T}\right) \tag{4.8}$$

With the above equations, the molar volume of a substance at P, T is (Shi et al. 1994):

$$V_{P,T} = V_{P_{ref},T}\left[1 + \frac{K' + K''\ln TP}{K_T}\right]^{-\left[K'+K''(T-T_{ref})\ln\left(\frac{T}{T_{ref}}\right)\right]^{-1}} \tag{4.9}$$

Because the isochoric heat capacity at $P = 10^5$ Pa (easily obtained from the isobaric counterpart by subtraction the anharmonicity term; Eq. (4.9)) must strictly conforms to the high-T Dulong-Petit limit (which is implicit in all computations based on first principles, but unfortunately almost never obeyed in thermodynamic assessments based on a collection of literature data) we may exert a fundamental control: heat capacity expansions that do not conform to the Dulong-Petit limit are unphysical, and the ensuing assessment must be simply rejected. We may utilize the Dulong-Petit constraint in substantially two different ways (Ottonello 2024a):

1. We accept the calorimetric data and tune the thermophysical parameters;
2. We accept the thermophysical parameters and tune the calorimetric observations.

The thermodynamic and thermophysical properties of all solids of interest in this study strictly conform to the above rules and are summarized in Tables 4.1 and 4.2.

## 4.2 Integration on the Intensive Variables

The effect of intensive variables T, P is accounted for in the usual way by Riemann's integration on the P–T limits of interest from T and P of reference ($T_r$, $P_r$):

$$\overline{H}_{T,P} = \overline{H}^0_{T_r,P_r} + \int_{T_r}^{T} \overline{C}_p dT + \int_{P_r}^{P}\left[\overline{V} - T\left(\frac{\partial V}{\partial T}\right)_P\right]dP \tag{4.10}$$

$$\overline{S}_{T,P} = \overline{S}^0_{T_r,P_r} + \int_{T_r}^{T} \overline{C}_{p_p}^{T} dT - \int_{P_r}^{P}\left(\frac{\partial V}{\partial T}\right)_P dP \tag{4.11}$$

$$\overline{G}_{T,P} = \overline{H}_{T,P} - T\overline{S}_{T,P} = \overline{H}^0_{T_r,P_r} - T\overline{S}^0_{T_r,P_r} + \int_{T_r}^{T} \overline{C}_p dT - \int_{T_r}^{T} \frac{\overline{C}_p}{T} dT + \int_{P_r}^{P} \overline{V}_r dP \tag{4.12}$$

with the horizontal hat stressing the fact that computations are carried out on a molar basis.

Based on the adopted heat capacity function, the integrals on T have the form:

$$\int_{T_r}^{T} C_p dT = a(T - T_r) + \frac{b}{2}(T^2 - T_r^2) - c(T^{-1} - T_r^{-1}) + 2d(T^{1/2} - T_r^{1/2})$$

**Table 4.1** Selected thermodynamic properties of solid phases in the CMAS system. $H^0_{f,298}$ = enthalpy of formation from the elements at $T_r = 298.15$ K, $P_r$ = 1 bar (integer); $S^0_{298}$ = entropy of the substance at the same T, P conditions; $C_P$ = isobaric heat capacity ($C_p = a + bT + cT^{-2} + dT^{-1/2} + eT^{-3} + fT^2 + gT^3 + hT^{-1}$)

| Composition (phase) | $H^0_{f,298.15}$ (J/mol) | $S^0_{298.15}$ (J/mol × K) | a | b × $10^3$ | c × $10^{-5}$ | d | e × $10^{-8}$ | f × $10^6$ | g × $10^9$ | h × $10^{-4}$ |
|---|---|---|---|---|---|---|---|---|---|---|
| α-$Al_2O_3$ (corundum) | −1,675,700 | 51.400 | 121.8 | 15.389 | −2.1403 | −3.2060 | −3.8874 | −3.5608 | 0.26997 | −0.88835 |
| γ-$Al_2O_3$ ($Fd\overline{3}m$) | −1,652,272 | 61.412 | 121.8 | 15.389 | −2.1403 | −3.2060 | −3.8874 | −3.5608 | 0.26997 | −0.88835 |
| $Al_2O_3$ (Pbca) | −1,641,000 | 62.800 | 80.36 | 37.0 | 0 | 0 | 0 | 0 | 0 | 0 |
| $Al_2O_3$ (perovskite) | −1,554,581 | 84.717 | 122.6 | 7.031 | −55.83 | 0 | 5.707 | 0 | 0 | −0.1404 |
| CaO (lime) | −645,725 | 37.750 | 66.329 | 0.6417 | 19.122 | −448.48 | −3.7626 | −0.4391 | 0.0489 | −0.133 |
| MgO (periclase) | −601,400 | 26.012 | 36.19 | 9.4773 | −23.853 | 271.93 | 2.2801 | −1.1724 | 0.1450 | 0 |
| $SiO_2$ (α-quartz) | −910,700 | 41.460 | 80.01 | 0 | −35.467 | −240.3 | 4.9157 | 0 | 0 | 0 |
| $SiO_2$ (β-quartz) | −908,627 | 44.207 | 80.01 | 0 | −35.467 | −240.3 | 4.9157 | 0 | 0 | 0 |
| $SiO_2$ (low-tridymite) | −907,610 | 44.029 | 75.37 | 0 | −59.581 | 0 | 9.5825 | 0 | 0 | 0 |
| $SiO_2$ (mean-tridymite) | −907,137 | 45.316 | 75.37 | 0 | −59.581 | 0 | 9.5825 | 0 | 0 | 0 |
| $SiO_2$ (high-tridymite) | −907,045 | 45.524 | 75.37 | 0 | −59.581 | 0 | 9.5825 | 0 | 0 | 0 |
| $SiO_2$ (α-cristobalite) | −907,753 | 43.394 | 83.51 | 0 | −24.554 | −374.7 | 2.8007 | 0 | 0 | 0 |
| $SiO_2$ (β-cristobalite) | −906,377 | 46.029 | 83.51 | 0 | −24.554 | −374.7 | 2.8007 | 0 | 0 | 0 |
| $SiO_2$ (coesite) | −906,602 | 40.668 | 97.552 | −2.3932 | −17.96 | −634.24 | 3.8529 | −0.65020 | 0.1238 | −0.2611 |
| $SiO_2$ (stishovite) | −874,361 | 27.809 | 73.768 | 1.0205 | −58.832 | 56.402 | 8.3752 | 0.6543 | 0 | 0 |
| $Al_2Si0_5$ (andalusite) | −2,589,972 | 91.434 | 236.48 | 0 | −75.268 | −1102.9 | 9.3644 | 0 | 0 | 0 |
| $Al_2Si0_5$ (kyanite) | −2,594,220 | 82.430 | 262.68 | 0 | −19.997 | −2001.4 | −0.6318 | 0 | 0 | 0 |
| $Al_2Si0_5$ (sillimanite) | −2,586,091 | 95.930 | 256.73 | 0 | −29.774 | −1887.2 | 2.5096 | 0 | 0 | 0 |
| $Al_6Si_20_{13}$ (stoich. mullite) | −6,820,400 | 274.889 | 634.81 | 0 | −172.099 | −3373.5 | 21.2274 | 0 | 0 | 0 |

(continued)

**Table 4.1** (continued)

| Composition (phase) | $H^0_{f,298.15}$ (J/mol) | $S^0_{298.15}$ (J/mol × K) | a | b × $10^3$ | c × $10^{-5}$ | d | e × $10^{-8}$ | f × $10^6$ | g × $10^9$ | h × $10^{-4}$ |
|---|---|---|---|---|---|---|---|---|---|---|
| $Ca_3SiO_5$ (hatrurite) | −2,929,202 | 168.599 | 321.19 | 0 | −9.948 | −2450.2 | 0.9753 | 0 | 0 | 0 |
| $Ca_2SiO_4$ (α-larnite) | −2,291,172 | 136.363 | 252.572 | 0 | 0 | −2182.967 | 0 | 0 | 0 | 0 |
| $Ca_2SiO_4$ (α′-bredigite) | −2,303,787 | 128.986 | 252.572 | 0 | 0 | −2182.967 | 0 | 0 | 0 | 0 |
| $Ca_2SiO_4$ (β-larnite) | −2,308,373 | 124.258 | 252.572 | 0 | 0 | −2182.967 | 0 | 0 | 0 | 0 |
| $Ca_2SiO_4$ (γ Ca-olivine) | −2,318,090 | 115.582 | 252.572 | 0 | 0 | −2182.967 | 0 | 0 | 0 | 0 |
| $Ca_3Si_20_7$ (rankinite) | −3,943,000 | 210.874 | 339.91 | 0 | −106.61 | −985.1 | 13.7359 | 0 | 0 | 0 |
| $CaSi0_3$ (wollastonite-I) | −1,632,094 | 81.810 | 149.07 | 0 | −36.593 | −690.3 | 4.8435 | 0 | 0 | 0 |
| $CaSiO_3$ (pseudo-wollast.) | −1,628,026 | 85.279 | 141.16 | 0 | −58.576 | −417.2 | 9.4074 | 0 | 0 | 0 |
| $CaSiO_3$ (wollastonite-II) | −1,626,762 | 82.100 | 128.51 | 4.2598 | −17.293 | 23.703 | −0.0028414 | 0.11076 | −0.829281 | 0 |
| $CaSiO_3$ (Ca-perovskite) | −1,541,000 | 74.000 | 114.66 | 9.8918 | −19.527 | −13.052 | 1.518 | −02.516 | −0.1538 | −0.47625 |
| $CaSi_2O_5$ (Ca-titanite) | −2,482,810 | 99.000 | 188.99 | 8.5971 | −56.16397 | 0 | 0 | 0 | 0 | 0 |
| $Ca_3Al_2O_6$ | −3,587,663 | 205.434 | 321.58 | 0 | −57.813 | −1255.4 | 6.6122 | 0 | 0 | 0 |
| $CaAl_2O_4$ | −2,326,229 | 114.223 | 227.04 | 0 | −5.608 | −1669.1 | −0.8554 | 0 | 0 | 0 |
| $CaAl_4O_7$ (grossite) | −3,999,090 | 177.820 | 337.98 | 0 | −121.126 | −1022.2 | 15.106 | 0 | 0 | 0 |
| $CaAl_{12}O_{19}$ (hibonite) | −10,704,186 | 388.910 | 992.735 | 0 | −281.478 | −5350.03 | 37.4701 | 0 | 0 | 0 |
| $MgAl_2O_4$ (spinel) | −2,300,313 | 84.535 | 235.9 | 0 | −17.104 | −1766 | 0.4062 | 0 | 0 | 0 |
| $MgSiO_3$ (*P*2/$_1$*c*) | −1,545,081 | 63.400 | 184.91 | −7.4034 | 29.272 | −1808.7 | −7.8593 | 0 | 0 | 0 |
| $MgSiO_3$ (*C*2/*c*) | −1,539,450 | 65.800 | 172.29 | −5.2079 | 8.7667 | −1429.3 | −4.5571 | 0 | 0 | 0 |
| $MgSiO_3$ (*Pbca*) | −1,545,552 | 66.170 | 133.75 | −0.89325 | −106.29 | 38.415 | 17.483 | 0 | 0 | 0 |

(continued)

**Table 4.1** (continued)

| Composition (phase) | $H^0_{f,298.15}$ (J/mol) | $S^0_{298.15}$ (J/mol × K) | a | b × $10^3$ | c × $10^{-5}$ | d | e × $10^{-8}$ | f × $10^6$ | g × $10^9$ | h × $10^{-4}$ |
|---|---|---|---|---|---|---|---|---|---|---|
| $MgSiO_3$ (*Pbcn*) | −1,544,308 | 67.438 | 132.86 | −0.7405 | −104.36 | 50.46 | 16.941 | 0 | 0 | 0 |
| $MgSiO_3$ (akimotoite) | −1,509,046 | 54.013 | 112.54 | 12.291 | −36.255 | 59.56 | 0 | −0.35901 | 0 | 0 |
| $MgSiO_3$ (majorite) | −1,516,134 | 68.506 | 183.7 | −17.173 | −9.1451 | −1507.3 | 0 | 4.6114 | 0 | 0 |
| $MgSiO_3$ (Mg-perovskite) | −1,457,192 | 58.698 | 110.12 | 9.5903 | −38.879 | 192.59 | 0 | 0.99651 | 0 | 0 |
| $Mg_2SiO_4$ (forsterite) | −2,173,000 | 94.100 | 155.13 | 21.307 | −44.06 | 87.536 | 0 | −0.7918 | 0 | 0 |
| $Mg_2SiO_4$ (wadsleyite) | −2,139,408 | 87.300 | 161.39 | 22.270 | −41.112 | −86.514 | 0 | −2.4731 | 0 | 0 |
| $Mg_2SiO_4$ (ringwoodite) | −2,126,979 | 81.700 | 153.46 | 21.405 | −47.49 | 127.08 | 0 | −3.08 | 0 | 0 |
| $Mg_{14}Si_5O_{24}$ (anhydrous phase B) | −13,180,000 | 561.200 | 975.27 | 111.77 | −271.02 | 352.26 | 0 | −3.9865 | 0 | 0 |
| $CaMgSiO_4$ (monticellite) | −2,250,027 | 108.300 | 226.34 | 0 | −11.797 | −1542.7 | −0.2329 | 0 | 0 | 0 |
| $CaMgSi_2O_6$ (*C2/c*) | −3,202,583 | 142.500 | 305.41333 | 0 | −71.659732 | −1604.931 | 9.2183754 | 0 | 0 | 0 |
| $CaMgSi_2O_6$ ($P2/_1c$) | −3,202,220 | 142.500 | 305.41333 | 0 | −71.659732 | −1604.931 | 9.2183754 | 0 | 0 | 0 |
| $CaMgSi_2O_6$ (*Pbca*) | −3,189,000 | 142.500 | 305.41333 | 0 | −71.659732 | −1604.931 | 9.2183754 | 0 | 0 | 0 |
| $CaMgSi_2O_6$ (*Pbcn*) | −3,185,512 | 142.500 | 305.41333 | 0 | −71.659732 | −1604.931 | 9.2183754 | 0 | 0 | 0 |
| $Ca_3MgSi_2O_8$ (merwinite) | −4,537,497 | 251.777 | 453.62 | 0 | 0 | −3250.0 | −3.4423 | −0 | 0 | 0 |
| $Ca_2MgSi_2O_{7n}$ (akermanite) | −3,862,527 | 212.000 | 387.06396 | 0 | 0 | −2938.8 | −0.4079 | 0 | 0 | 0 |
| $Mg_2Al_4Si_5O_{18}$ (cordierite) | −9,193,385 | 384.620 | 954.39 | 0 | −23.173 | −7962.3 | −3.7021 | 0 | 0 | 0 |
| $Mg_3Al_{10}SiO_{20}$ (sapphirine-351) | −11,199,294 | 397.575 | 931.33 | 52.038 | 105.44 | −4974.6 | −62.714 | −9.4942 | 1.4319 | 0 |
| $Mg_4Al_8Si_2O_{20}$ (sapphirine-442) | −11,005,000 | 406.975 | 886.82 | 67.957 | 57.970 | −3846.2 | −55.243 | −12.915 | 1.7113 | 0 |

(continued)

**Table 4.1** (continued)

| Composition (phase) | $H^0_{f,298.15}$ (J/mol) | $S^0_{298.15}$ (J/mol × K) | a | b × $10^3$ | c × $10^{-5}$ | d | e × $10^{-8}$ | f × $10^6$ | g × $10^9$ | h × $10^{-4}$ |
|---|---|---|---|---|---|---|---|---|---|---|
| $Mg_3Al_2Si_3O_{12}$ (pyrope) | −6,297,697 | 254.479 | 631.5 | 0 | 0 | −4462 | −12.605 | 0 | 0 | 0 |
| $CaAl_2Si_2O_8$ (anorthite) | −4,232,730 | 200.186 | 439.37 | 0 | 0 | −3734.1 | −3.1302 | 0 | 0 | 0 |
| $Ca_2Al_2SiO_7$ (gehlenite-a) | −3,988,158 | 198.600 | 373.0874 | 0 | −47.78466 | −2276.755 | 4.779118 | 0 | 0 | 0 |
| $Ca_2Al_2SiO_7$ (gehlenite-b) | −3,990,951 | 189.215 | 151.3474 | 369.5 | 125.12814 | −2276.755 | 4.779118 | −146.9 | 0 | 0 |
| $Ca_2Al_2SiO_7$ (gehlenite-c) | −3,975,168 | 210.068 | 373.0874 | 0 | −47.78466 | −2276.755 | 4.779118 | 0 | 0 | 0 |
| $Ca_3Al_2Si_3O_{12}$ (grossular) | −6,632,859 | 255.150 | 573.43 | 0 | −188.872 | −2039.4 | 23.1931 | 0 | 0 | 0 |
| $CaAl_2SiO_6$ (Ca-Tschermak) | −3,296,800 | 144.500 | 310.7 | 0 | −74.553 | −1671.6 | 9.4878 | 0 | 0 | 0 |
| $Ca_3MgAl_4O_{10}$ | −5,899,975 | 290.087 | 631.86 | −7.9146 | 24.123 | −5312.16 | 0 | 0 | 0 | 0 |
| $Ca_2Mg_2Al_{28}O_{46}$ | −26,014,050 | 897.124 | 2388.6 | −20.5686 | −2428.38 | −7227 | 670.86 | 0 | 0 | 0 |
| $CaMg_2Al_{16}O_{27}$ | −15,197,125 | 600.974 | 1706.21 | −37.9115 | 81.4495 | −16,578.1 | 0 | 0 | 0 | 0 |

**Table 4.2** Selected thermophysical properties of solid phases in the CMAS system. $V^0_{298.15}$ = standard-state molar volume ($T_r$ = 298.15 K, $P_r$ = 1 bar). $K_0$ = bulk modulus at $T_r$ = 298.15 K and $P_r$ = 1 bar; $K'_0 = (dK/dP)_T$ = pressure derivative of the bulk modulus at $T_r$ = 298.15 K and $P_r$ = 1 bar; $K_1$ and $K^2$ = thermal dependence of compressibility; $(dK'/dT)_P$ = temperature derivative of the bulk modulus pressure derivative; $\alpha_V = \alpha_0 T + \alpha_1 + \alpha_2 T^{-1} + \alpha_3 T^{-2} + \alpha_4 T^{-3}$ = thermal expansion coefficient at T, $P_r$. $T_{max}$ is the maximum temperature of validity for the adopted thermophysical properties

| Composition (phase) | $V^0_{298.15}$ (cc/mol) | $K_0$ (GPa) | $K'$ | $K_1$ (bar/K) | $K_2$ (bar/$K^2$) | $K''$ ($K^{-1}$) | $\alpha_0 \times 10^7$ ($K^{-2}$) | $\alpha_1 \times 10^7$ ($K^{-1}$) | $\alpha_2 \times 10^3$ | $\alpha_3$ (K) | $\alpha_4$ ($K^2$) |
|---|---|---|---|---|---|---|---|---|---|---|---|
| α-$Al_2O_3$ (corundum) | 25.575 | 266.6 | 4.2 | −350 | 0 | 0 | −0.008 | 419.6 | −11.406 | 1.0623 | 0 |
| γ-$Al_2O_3$ ($Fd\bar{3}m$) | 36.414 | 266.6 | 4.2 | −350 | 0 | 0 | −0.008 | 419.6 | −11.406 | 1.0623 | 0 |
| $Al_2O_3$ (Pbca) | 26.78 | 225 | 4.2 | −50 | 0 | 0 | 0.12805 | 203.03 | 0 | 0 | 0 |
| $Al_2O_3$ (perovskite) | 25.3 | 276.7 | 4.1 | −390.9 | 0 | 0 | 0.045677 | 4 | 80.2 | −8.5134 | 0.11908 |
| CaO (lime) | 16.76 | 130 | 4 | −320.904 | 0.0251 | 0 | 0 | 501.25 | −5.561 | −0.4337 | 0 |
| MgO (periclase) | 11.25 | 159.5 | 4.04 | −296.9 | 0.0205 | 0 | 0.1183664 | 301.3735 | 4.49044 | −2.50218 | 241.71591 |
| $SiO_2$ (α-quartz) | 22.69 | 48 | 6 | −405.5 | 0.2126 | 0 | −0.6756 | 699.08 | 0 | 0 | 0 |
| $SiO_2$ (β-quartz) | 23.9 | 132.6 | 3.9 | −1153.8 | 0.3218 | 0 | −0.07 | 0 | 0 | 0 | 0 |
| $SiO_2$ (low-tridymite) | 27.17 | 30.8 | 4.1 | 0 | 0 | 0 | −0.27655 | 313.204 | 0 | 0 | 0 |
| $SiO_2$ (mean-tridymite) | 27.17 | 30.8 | 4.1 | 0 | 0 | 0 | −0.27655 | 313.204 | 0 | 0 | 0 |
| $SiO_2$ (high-tridymite) | 27.306 | 30.8 | 4.1 | 0 | 0 | 0 | −0.27655 | 313.204 | 0 | 0 | 0 |
| $SiO_2$ (α-cristobalite) | 25.76 | 11.5 | 9 | 0 | 0 | 0 | 2.55 | −2424.5 | 0 | 0 | 0 |
| $SiO_2$ (β-cristobalite) | 27.242 | 14 | 7.1 | −20 | 0 | 0 | −0.173593 | 235.599 | 0 | 0 | 0 |
| $SiO_2$ (coesite) | 20.64 | 100.3 | 8.4 | −249.584 | 0.0252 | 0 | −0.004534 | 101.67 | 0 | 0 | 0 |
| $SiO_2$ (stishovite) | 14 | 270.1 | 6.59 | −474.75 | 0.0351 | 0 | 0.05904 | 167.77 | 2.2872 | −2.1518 | 237.0245 |
| $Al_2SiO_5$ (andalusite) | 51.47 | 139 | 4 | −78 | 0 | 0 | 0.007161 | 233.6 | 0 | 0 | 0 |
| $Al_2SiO_5$ (kyanite) | 44.12 | 193 | 4 | 153 | −0.051 | 0 | 0.009468 | 270.92 | −0 | 0 | 0 |
| $Al_2SiO_5$ (sillimanite) | 49.83 | 171 | 4 | −257 | 0 | 0 | 0.0071 | 140.7 | 0 | 0 | 0 |

(continued)

**Table 4.2** (continued)

| Composition (phase) | $V^0_{298.15}$ (cc/mol) | $K_0$ (GPa) | K′ | $K_1$ (bar/K) | $K_2$ (bar/$K^2$) | K″ ($K^{-1}$) | $\alpha_0 \times 10^7$ ($K^{-2}$) | $\alpha_1 \times 10^7$ ($K^{-1}$) | $\alpha_2 \times 10^3$ | $\alpha_3$ (K) | $\alpha_4$ ($K^2$) |
|---|---|---|---|---|---|---|---|---|---|---|---|
| $Al_6Si_20_{13}$ (stoich. mullite) | 134.6 | 168.3 | 4 | −175 | 0 | 0 | 0 | 167 | 0 | 0 | 0 |
| $Ca_3SiO_5$ (hatrurite) | 72.74 | 97 | 4 | −263.6 | 0.023 | 0 | −0.0325 | 956.74 | −20.0261 | 0 | 0 |
| $Ca_2SiO_4$ (α-larnite) | 53.6 | 125 | 4 | −290 | 0.018 | 0 | 0.047 | 508.93 | −5.412 | 0 | 0 |
| $Ca_2SiO_4$ (α′-bredigite) | 51.8 | 125 | 4 | −290 | 0.018 | 0 | 0.047 | 508.93 | −5.412 | 0 | 0 |
| $Ca_2SiO_4$ (β-larnite) | 51.6 | 125 | 4 | −290 | 0.018 | 0 | 0.047 | 508.93 | −5.412 | 0 | 0 |
| $Ca_2SiO_4$ (γ Ca-olivine) | 57.5 | 122 | 4 | −290 | 0.018 | 0 | 0.047 | 508.93 | −5.412 | 0 | 0 |
| $Ca_3Si_20_7$ (rankinite) | 96.51 | 100.5 | 4 | −100 | 0 | 0 | −0.0028 | 363.56 | −0.874 | 0 | 0 |
| $CaSi0_3$ (wollastonite-I) | 39.62 | 116.6 | 4 | −330 | 0.027 | 0 | 0.04664 | 339.404 | 1.4588 | −1.8634 | 0 |
| $CaSiO_3$ (pseudo-wollaston.) | 40.08 | 113 | 4 | −300 | 0.029 | 0 | −0.013 | 466 | −8.3137 | 0 | 0 |
| $CaSi0_3$ (wollastonite-II) | 37.94 | 93.9 | 5 | 0 | 0 | 0 | 0.01998 | 320 | 0 | 0 | 0 |
| $CaSi0_3$ (Ca-perovskite) | 27.45 | 239 | 4.8 | −300 | 0.01 | 0 | 0.01 | 300 | 0 | 0 | 0 |
| $CaSi_20_5$ (Ca-titanite) | 48.192 | 178.2 | 4 | −376 | 0.039 | 0 | 0 | 350 | 0 | 0 | 0 |
| $Ca_3Al_2O_6$ | 88.955 | 102.9 | 4 | −292.2 | 0.0225 | 0 | 0.07564 | 255.5 | 0 | −0.749 | 0 |
| $CaAl_2O_4$ | 39.76 | 219 | 4 | 103.9 | −0.048 | 0 | 0 | 280 | 0 | 0 | 0 |
| $CaAl_4O_7$ (grossite) | 89.057 | 200 | 4 | −466.7 | 0.036 | 0 | 0 | 275.4 | 0 | 0 | 0 |
| $CaAl_{12}O_{19}$ (hibonite) | 176.729 | 230 | 4 | 190 | −0.07 | 0 | 0 | 280 | 0 | 0 | 0 |
| $MgAl_2O_4$ (spinel) | 39.77 | 217.9 | 4.89 | −561.994 | 0.0452 | 0.0009 | 0.01447 | 500.8 | −11.0 | 0 | 0 |
| $MgSiO_3$ ($P2/_1c$) | 31.255 | 121.4 | 5.5 | −210 | 0 | 0 | 0.007956 | 300.1 | −1.599 | 0 | 0 |

(continued)

**Table 4.2** (continued)

| Composition (phase) | $V^0_{298.15}$ (cc/mol) | $K_0$ (GPa) | $K'$ | $K_1$ (bar/K) | $K_2$ (bar/$K^2$) | $K''$ ($K^{-1}$) | $\alpha_0 \times 10^7$ ($K^{-2}$) | $\alpha_1 \times 10^7$ ($K^{-1}$) | $\alpha_2 \times 10^3$ | $\alpha_3$ (K) | $\alpha_4$ ($K^2$) |
|---|---|---|---|---|---|---|---|---|---|---|---|
| $MgSiO_3$ (*C2/c*) | 30.337 | 126.3 | 5.5 | −184 | 0 | 0 | 0.006784 | 303.3 | −1.7823 | 0 | 0 |
| $MgSiO_3$ (*Pbca*) | 31.33 | 105.8 | 8.5 | −206 | 0 | 0.0015 | 0.02694 | 294.7 | 0 | 0 | −0.5588 |
| $MgSiO_3$ (*Pbcn*) | 32.42 | 107 | 6.6 | −200 | 0 | 0 | 0.03 | 266 | 0 | 0 | 0 |
| $MgSiO_3$ (akimotoite) | 26.354 | 207.7 | 4.54 | −279 | 0 | 0 | 0.10997 | 82.956456 | 17.990566 | −8.0792 | 1019.1 |
| $MgSiO_3$ (majorite) | 28.573 | 160.1 | 4.82 | −237 | 0 | 0 | 0.1361365 | 24.239955 | 23.562897 | −10.0127 | 1287.32 |
| $MgSiO_3$ (Mg-perovskite) | 24.318 | 249.2 | 4.2 | −280 | 0 | 0 | 0.0818 | 198.2 | 0 | −0.474 | 0 |
| $Mg_2SiO_4$ (forsterite) | 43.67 | 130.8 | 4 | −174.6 | 0 | 0 | 0.10031 | 269.31 | 5.6082 | −2.6162 | 243.828 |
| $Mg_2SiO_4$ (wadsleyite) | 40.167 | 161.8 | 4.415 | −118 | 0 | 0 | 0 | 464 | −13.8005 | 3.6196 | −395.094 |
| $Mg_2SiO_4$ (ringwoodite) | 38.933 | 196.4 | 4.322 | −104.1 | 0 | 0 | 0 | 323 | −7.3084 | 1.3745 | −150.406 |
| $Mg_{14}Si_5O_{24}$ (anhydrous phase B) | 251.73 | 155 | 4.14 | −189 | 0 | 0 | 0.08628 | 230.86 | 8.313 | −4.3445 | 528.465 |
| $CaMgSiO_4$ (monticellite) | 51.48 | 114 | 4 | −301.8 | 0.0206 | 0 | 0.09496 | 292.5 | 0 | 0 | 0 |
| $CaMgSi_2O_6$ (*C2c* diopside) | 66.2 | 126.7 | 4 | −382.6 | 0.0313 | 0.0005 | 0.098 | 251 | 5.191 | −2.365 | 0 |
| $CaMgSi_2O_6$ (*P2/*$_1$*c*) | 66.2 | 118 | 4 | −192 | 0 | 0.0005 | 0.098 | 251 | 5.191 | −2.365 | 0 |
| $CaMgSi_2O_6$ (*Pbca*) | 66.2 | 195 | 4.2 | −315 | 0 | 0.0001 | 0.098 | 251 | 5.191 | −2.365 | 0 |
| $CaMgSi_2O_6$ (*Pbcn*) | 66.2 | 116 | 4 | −184 | 0 | 0.0005 | 0.098 | 251 | 5.191 | −2.365 | 0 |
| $Ca_3MgSi_2O_8$ (merwinite) | 98.47 | 189 | 4 | −663 | 0.066 | 0 | 0.04857 | 403.5 | −0.03715 | 0 | 0 |
| $Ca_2MgSi_2O_7$ (akermanite) | 92.52 | 93.5 | 4 | 0 | 0 | 0 | 0 | 321 | 0 | 0 | 0 |

(continued)

**Table 4.2** (continued)

| Composition (phase) | $V^0_{298.15}$ (cc/mol) | $K_0$ (GPa) | $K'$ | $K_1$ (bar/K) | $K_2$ (bar/$K^2$) | $K''$ ($K^{-1}$) | $\alpha_0 \times 10^7$ ($K^{-2}$) | $\alpha_1 \times 10^7$ ($K^{-1}$) | $\alpha_2 \times 10^3$ | $\alpha_3$ (K) | $\alpha_4$ ($K^2$) |
|---|---|---|---|---|---|---|---|---|---|---|---|
| $Mg_2Al_4Si_5O_{18}$ (cordierite) | 233.11 | 128 | 4 | −153 | 0 | 0 | 0.01871 | 377.1 | −24.46 | 4.048 | 0 |
| $Mg_3Al_{10}SiO_{20}$ (sapphirine-351) | 202.387 | 172.3 | 4.28 | −188 | 0 | 0 | 0.060817 | 157.09 | 8.283 | −4.3554 | 547.232 |
| $Mg_4Al_8Si_2O_{20}$ (sapphirine-442) | 204.879 | 158.7 | 4.68 | −210.2 | 0 | 0 | 0.090302 | 104.11 | 14.034 | −6.4469 | 814.288 |
| $Mg_3Al_2Si_3O_{12}$ (pyrope) | 113.16 | 166 | 4 | −225 | 0 | 0 | 0.02886 | 287.4 | 0 | −0.5443 | 0 |
| $CaAl_2Si_2O_8$ (anorthite) | 100.75 | 100 | 4 | −256.4 | 0.017 | 0 | 0.12106 | 234.041 | 5.20652 | −2.19267 | 0 |
| $Ca_2Al_2SiO_7$ (gehlenite-a) | 90.08 | 106.4 | 4 | −2.3 | −0.0253 | 0 | 0.04323 | 217.7 | 0 | 0 | 0 |
| $Ca_2Al_2SiO_7$ (gehlenite-b) | 90.08 | 106.4 | 4 | −2.3 | −0.0253 | 0 | 0.04323 | 217.7 | 0 | 0 | 0 |
| $Ca_2Al_2SiO_7$ (gehlenite-c) | 90.08 | 106.4 | 4 | −2.3 | −0.0253 | 0 | 0.04323 | 217.7 | 0 | 0 | 0 |
| $Ca_3Al_2Si_3O_{12}$ (grossular) | 125.38 | 167.3 | 4.887 | −460 | 0.035 | 0 | 0.05272 | 222.4 | 0 | −0.4119 | 0 |
| $CaAl_2SiO_6$ (Ca-Tschermak) | 63.56 | 124.1 | 4.2 | −333.3 | 0.0269 | 0 | 0.047557 | 361.989 | 8.7283 | −4.0227 | 0 |
| $Ca_3MgAl_4O_{10}$ | 138.9022 | 200 | 4 | −200 | 0 | 0 | 0 | 310 | 0 | 0 | 0 |
| $Ca_2Mg_2Al_{28}O_{46}$ | 2146.5 | 200 | 4 | −200 | 0 | 0 | 0 | 150 | 0 | 0 | 0 |
| $CaMg_2Al_{16}O_{27}$ | 846.8 | 200 | 4 | −200 | 0 | 0 | 0 | 180 | 0 | 0 | 0 |

$$- \frac{e}{2}\left(T^{-2} - T_r^{-2}\right) + \frac{f}{3}\left(T^3 - T_r^3\right) + \frac{g}{4}\left(T^4 - T_r^4\right) + h\ln\left(\frac{T}{T_r}\right) \quad (4.13)$$

$$\int_{T_r}^{T} \frac{C_p}{T} dT = a\ln\left(\frac{T}{T_r}\right) + b(T - T_r) - \frac{c}{2}\left(T^{-2} - T_r^{-2}\right) - 2d\left(T^{-1/2} - T_r^{-1/2}\right) - \frac{e}{3}\left(T^{-3} - T_r^{-3}\right) + \frac{f}{2}\left(T^2 - T_r^2\right) + \frac{g}{3}\left(T^3 - T_r^3\right) - h\left(T^{-1} - T_r^{-1}\right) \quad (4.14)$$

With a third-order Birch-Murnaghan equation of state and adopting the procedure of Ganguly (2008), one has:

$$\int_{P_{ref}}^{P'} VdP = \int_{P_rV(P_r)}^{P'V(P')} d(PV) - \int_{V_r}^{V'} PdV = \left[P'V(P') - P_{ref}V(P_{ref})\right] - \int_{V_r}^{V'} PdV \quad (4.15)$$

$$\int_{V_r}^{V'} PdV = \frac{9}{4}K_0V_0^{5/3}\left[\frac{1}{V^{2/3}} - \frac{1}{2V^{4/3}}\right]_{V_r}^{V'} + \frac{9}{16}K_0\left(K'' - 4\right)\left[\frac{1}{V^2} + \frac{3}{2V_0^{2/3}V^{4/3}} - \frac{1}{3V_0^{4/3}V^{2/3}}\right]_{V_r}^{V'} \quad (4.16)$$

The above formalism may result in physical inconsistencies when applied at very high P, T conditions, as discussed in Helffrich and Connolly (2009), but has no practical drawbacks in the present context. Adopting for V(P, T) Eq. (4.9) Shi et al. (1994), $P_{ref} = 1$, $T_{ref} = 298.15$ obtain:

$$\int_{P_{ref}}^{P} V(P,T)dP = V_{P_{ref},T_{ref}}\left[\begin{array}{c} 1 + \alpha_0(T - T_{ref}) + \frac{\alpha_1}{2}\left(T^2 - T_{ref}^2\right) \\ + \infty\alpha_2\ln\left(\frac{T}{T_{ref}}\right) - \alpha_3\left(\frac{1}{T} - \frac{1}{T_{ref}}\right) \end{array}\right] \times \left\{\frac{K_T}{K' + K''(T - T_{ref})\ln\left(\frac{T}{T_{ref}}\right) - 1}\left[\left(1 + \frac{K' + K''(T - T_{ref})\ln\frac{T}{T_{ref}}}{\cdots}\times\right)^B - 1\right)\right\} \quad (4.17)$$

$$B = \frac{K' + K''(T - T_{ref})\ln\left(\frac{T}{T_{ref}}\right) - 1}{K' + K''(T - T_{ref})\ln\left(\frac{T}{T_{ref}}\right)} \quad (4.18)$$

The simplest procedure exploits the benefits of modern computing and explores in an iterative fashion the value of V(P, T) ($V_{trial}$) consistent with the adopted EoS, i.e., for a third-order Birch-Murnaghan EoS:

$$P = P_{trial} = \frac{3}{2}K_T\left[\left(\frac{V_0}{V_{trial}}\right)^{7/3} - \left(\frac{V_0}{V}\right)^{5/3}\right]$$
$$\left\{1 - \frac{3}{4}(4 - K'')\left[\left(\frac{V_0}{V_{trial}}\right)^{2/3} - 1\right]\right\} + P_0 \tag{4.19}$$

Having the appropriate value of V(P, T) at any condition of interest, the VdP integral is readily obtained by Gaussian quadrature. The various formulations agree at the level of less than 1 J on a molar unit at the maximum P, T conditions of interest, i.e., 5000 K, 136 GPa.

Because for some solids we adopted $K''$ consistent with the formulation of Shi et al. (1994), we adopt in this study Eqs. (4.23) and (4.24).

## 4.3 Mechanical Mixtures

In the investigated P, T range ($10^5$ to ~$10^{11}$ Pa; 0–5 × $10^3$ K), we encounter in the CMAS system 66 solids which, with reasonable approximation, may be treated as pure components. Their (refined) thermodynamic and thermophysical parameters are collected in Tables 4.1 and 4.2.

## 4.4 Polymorphic Transitions

Several of the pure solids undergo first or second-order (lambda) polymorphic transitions. Lambda transitions are treated according to the procedure of Berman and Brown (1985, 1986), Berman et al. (1985), Berman (1988) by the following equation (applied for $T_{ref,\lambda} < T < T_\lambda$ at 1-bar pressure):

$$C_{P,\lambda} = T \cdot (X_{\lambda 1} + X_{\lambda 2}T)^2 = b_\lambda T + f_\lambda T^2 + g_\lambda T^3 \tag{4.20}$$

where the parameters $b_\lambda$, $f_\lambda$, and $g_\lambda$ obey the functional form of the Shomate Eq. (4.1).

The effect of P on the lambda transition temperature $T_\lambda$ is assumed to be strictly linear (Berman 1988):

$$T_\lambda^P = T_\lambda^{1bar} + k(P - 1)) \tag{4.21}$$

Then, posing (Berman 1988):

$$T_d = T_\lambda^{1bar} - T_\lambda^P \tag{4.22}$$

$$T' = T + T_d \tag{4.23}$$

At any pressure, for $T_{ref} < T' < T_\lambda^{1bar}$ the heat capacity of the lambda transition (to be added to the normal or "*lattice*" heat capacity (Berman 1988) is

$$C_{p,\lambda}^P = T'(X_{\lambda 1} + X_{\lambda 2}T')^2 \tag{4.24}$$

Posing

$$t_r = T_{ref} - T_d \tag{4.25}$$

$$x_1 = X_{\lambda 1}^2 T_d + 2X_{\lambda 1}X_{\lambda 2}T_d^2 + X_{\lambda 2}^2 T_d^3 \tag{4.26}$$

$$x_2 = X_{\lambda 1}^2 + 4X_{\lambda 1}X_{\lambda 2}T_d + 2X_{\lambda 2}^2 T_d^2 \tag{4.27}$$

$$x_3 = 2X_{\lambda 1}X_{\lambda 2}T_d + 3X_{\lambda 2}^2 T_d^2 \tag{4.28}$$

$$x_4 = X_{\lambda 2}^2 \tag{4.29}$$

The lambda contributions to the lattice enthalpy and entropy are respectively given by (Berman 1988):

$$\Delta_\lambda H^{P,T} = \int_{t_r}^{T} C_{p,\lambda}^P dT = x_1(T - t_r) + \frac{x_2}{2}(T^2 - t_r^2) + \frac{x_3}{3}(T^3 - t_r^3) + \frac{x_4}{4}(T^4 - t_r^4) \tag{4.30}$$

$$\Delta_\lambda S^{P,T} = \int_{t_r}^{T} C_{p,\lambda}^P \frac{dT}{T} = x_1 \ln\left(\frac{T}{t_r}\right) + x_2(T - t_r) + \frac{x_3}{2}(T^2 - t_r^2) + \frac{x_4}{4}(T^3 - t_r^3) \tag{4.31}$$

$$\Delta_\lambda G^{P,T} = \Delta_\lambda H^{P,T} - T\Delta_\lambda S^{P,T} \tag{4.32}$$

In the case of mixed transitions (1st + 2nd order), the first-order contributions must be added as well (see Table 4.3).

**Table 4.3** Polymorphic transitions (Lambda and first-order contributions)

| Low-T polymorph | $T_\lambda$ | $T_{ref}$ | k | $X_{\lambda 1}$ | $X_{\lambda 2}$ | $B_\lambda$ | $F_\lambda$ | $G_\lambda$ | $\Delta H_{1st\ order}$ | Refs. |
|---|---|---|---|---|---|---|---|---|---|---|
| $SiO_2$ (α-quartz) | 848 | 373 | 0.02565 | −0.09187 | 0.00024607 | 0.008440 | −4.52129E-05 | 6.05504E-08 | 499 | (1) |
| $SiO_2$ (α-cristobalite) | 550 | 298.15 | 0.1217 | −0.1 | 0.0004465 | 0.010000 | −8.000000E-5 | 1.60000E-07 | 304 | (2) |
| $SiO_2$ (low-tridymite) | 388 | 304 | 0.0 | −0.83 | 0.0027311 | 0.705600 | −4.586000E-3 | 7.452900E-6 | – | (2) |
| $SiO_2$ (mean-trydimite) | 440 | 388 | 0.0655 | −0.52 | 0.0014 | 0.270400 | −1.456000E-3 | 1.96000E-06 | – | (2) |
| α′-$Ca_2SiO_4$ | 1710 | 970 | 0.023 | −0.22815 | 0.00023196 | 0.052052 | −1.058400E-4 | 5.38054E-08 | 12,615 | (2) |

(1) Berman and Brown (1985); (2) Belmonte et al. (2017b)

## 4.5 Solid Mixtures

Several of the solids nucleating in the CMAS system are 2-components mixtures. Their solid mixing behavior is almost never ideal, and the deviation from ideality is conveniently reproduced through multiparametric Redlich–Kister expansions of Van Laar coefficients (Redlich and Kister 1948), as follows:

$$G_{excess} = X_1X_2\left[A_{0G} + A_{1G}(X_1 - X_2) + A_{2G}(X_1 - X_2)^2\right] \tag{4.33}$$

with:

$$A_{nG} = A_{nH} - TA_{nS} + PA_{nV} \tag{4.34}$$

The adopted parameters are summarized in Table 4.4. In most cases, the thermodynamic and thermophysical properties of components in a mixture are those listed in Tables 4.1 and 4.2 and the excess parameters were obtained from a literature survey. Generally, we do not have enough experimental information to confirm solid mixing properties at pressure, but the topological analysis tells us that the univariant equilibria and invariant points at room pressure are, in most cases, not much affected by mixing at the solid state. Some solid mixtures deserve some extended comment, nevertheless.

### *4.5.1 Mullite*

Mullite is structurally very similar to sillimanite (Schneider and Komarneni 2005; Schneider et al. 2008) and, according to Hillert et al. (1989), the $G^0{}_f$ of stoichiometric mullite ($Al_6Si_2O_{13}$) is nearly equivalent to the summation of the chemical potentials of the end-members plus an ideal mixing contribution on the third sublattice (i.e., $Al_IAl_{II}[Al_{0.25}Si_{0.75}]_{III}[O_{0.975}\square_{0.025}]_{5,IV}$, where I, II, III, and IV indicate four different sub-lattices in the structure). With Howald and Eliezer (1978), we did not modify the energy properties of the two end-member components, though they are not isostructural. We accepted that mixing takes place only in the third sublattice since vacancies are strictly associated with $Al^{3+}$, having thus no configurational degrees of freedom, but included a slight excess term. The assessed Redlich–Kister parameters for the mullite solid mixture are purely entropic, i.e., $A_{0S} = -16$ J/(mol × K); $A_{3S} = -15$ J/(mol × K (Table 4.4). We do not have enough experimental data to confirm the solid mixture at pressure.

**Table 4.4** Thermodynamic and excess mixing properties of solid mixtures in the CMAS system

| Solid mixture | Spinel | | Sapphirine | | Orthopyroxene | | Perovskite | | Mullite | | Oxide | | Garnet | |
|---|---|---|---|---|---|---|---|---|---|---|---|---|---|---|
| End-member | $MgAl_2O_4$ | γ-$AlAlO_3$ | $Mg_3Al_{10}SiO_{20}$ | $Mg_4Al_8Si_2O_{20}$ | $MgSiO_3$ | *Pbca* $AlAlO_3$ | $MgSiO_3$ | $AlAlO_3$ | $SiO_2$ | $Al_2O_3$ | MgO | $Al_2O_3$ | $Mg_3Al_2Si_3O_{12}$ | $Mg_4Si_4O_{12}$ |
| $H^0_{f,298.15}$ (kJ/mol) | −2,300,313 | −1,652,271.5 | −11,199,294 | −11,005,000 | −1,545,552 | −1,641,000 | −1,457,192 | −1,554,581 | −906,377 | −1,675,700 | −601,400 | −1,598,296 | −6,297,697 | −6,064,536 |
| $S^0_{298.15}$ [J/(mol × K)] | 84.535 | 61.4125 | 397.575 | 406.975 | 66.17 | 62.8 | 58.698 | 84.717 | 46.029 | 51.4 | 26.012 | 51.4 | 254.479 | 274.024 |
| a | 235.9 | 121.8 | 931.33 | 886.82 | 133.75 | 80.36 | 110.120 | 122.600 | 83.51 | 121.8 | 36.190 | 121.8 | 631.5 | 734.8 |
| $b \times 10^3$ | 0.000 | 15.389 | 52.038 | 67.957 | −0.89325 | 37 | 9.5903 | 7.031 | 0.000 | 15.389 | 9.4773 | 15.389 | 0 | −68.692 |
| $c \times 10^{-5}$ | −17.104 | −2.1403 | 105.44 | 57.970 | −106.029 | 0 | −38.879 | −55.830 | −24.554 | −2.1403 | −23.853 | −2.1403 | 0 | −36.5804 |
| d | −1766.0 | −3.2060 | −4974.6 | −3846.2 | 38.415 | 0 | 192.590 | 0.000 | −374.7 | -3.206 | 271.930 | -3.206 | −4462 | −6029.2 |
| $e \times 10^{-8}$ | 0.4062 | −3.8874 | −62.714 | −55.243 | 17.483 | 0 | 0.000 | 5.707 | 2.8007 | −3.8874 | 2.2801 | −3.8874 | −12.605 | 0 |
| $f \times 10^6$ | 0.000 | −3.5608 | −9.4942 | −12.915 | 0 | 0 | 0.99651 | 0.000 | 0.000 | −3.5069 | −1.1724 | −3.5068 | 0.000 | 18.4456 |
| $g \times 10^9$ | 0.000 | 0.26997 | 1.4319 | 1.7113 | 0 | 0 | 0.000 | 0.000 | 0.000 | 0.26997 | 0.14495 | 0.26997 | 0 | 0.000 |
| $h \times 10^{-4}$ | 0.000 | −0.88835 | 0.000 | 0.000 | 0 | 0 | 0.000 | −0.1404 | 0.000 | −0.88835 | 0.000 | −0.88835 | 0.000 | 0 |
| $V^0_{298.15}$ (cc/mol) | 39.770 | 36.414 | 202.39 | 204.879 | 31.330 | 26.78 | 24.318 | 25.300 | 27.242 | 25.575 | 11.250 | 25.575 | 113.16 | 114.292 |
| $A_{0,H}$ (kJ/mol) | −17.6 | – | 0 | – | 0 | – | 0 | – | 0 | – | 42.9 | – | 0 | – |
| $A_{0,S}$ [J/(mol × K)] | −25 | – | 0 | – | −10 | – | 0 | – | −14 | – | 0 | – | 0 | – |
| $A_{0,V}$ (J/bar) | 0 | – | 0 | – | 0 | – | 0 | – | 0 | – | 0 | – | 0 | – |
| $A_{3,H}$ (kJ/mol) | 0 | – | 0 | – | 0 | – | 0 | – | 0 | – | 0 | – | 0 | – |
| $A_{3,S}$ [J/(mol × K)] | 0 | – | 0 | – | 0 | – | 0 | – | −16 | – | 0 | – | 0 | – |
| $A_{3,V}$ (J/bar) | 0 | – | 0 | – | 0 | – | 0 | – | 0 | – | 0 | – | 0 | – |
| Nixing sites | 1 | | 1 | | 2 | | 2 | | 1 | | 1 | | 2 | |

(continued)

**Table 4.4** (continued)

| Solid mixture | Garnet | | Oxide | | Orthopyroxene (*Pbca*) | | High-clinopyroxene (*C2/c*) | | Low-clinopyroxene ($P2/_1c$) | | Protopyroxene (*Pbcn*) | |
|---|---|---|---|---|---|---|---|---|---|---|---|---|
| End-member | $Mg_3Al_2Si_3O_{12}$ | $Ca_3Al_2Si_3O_{12}$ | MgO | CaO | $CaMgSi_2O_6$ | $Mg_2Si_2O_6$ | $CaMgSi_2O_6$ | $Mg_2Si_2O_6$ | $CaMgSi_2O_6$ | $Mg_2Si_2O_6$ | $CaMgSi_2O_6$ | $Mg_2Si_2O_6$ |
| $H^0_{f,298.15}$ (kJ/mol) | −6,297,697 | −6,632,859 | −601,400 | −645,725 | −3,189,000 | −3,091,104 | −3,202,583 | −3,078,900 | −3,202,220 | −3,090,162 | −3,185,511.8 | −3,088,616 |
| $S^0_{298.15}$ [J/(mol × K)] | 254.479 | 255.15 | 26.012 | 37.75 | 142.50 | 132.34 | 142.5 | 131.6 | 142.50 | 126.8 | 142.50 | 134.876 |
| a | 631.5 | 573.43 | 36.190 | 66.329 | 305.41333 | 267.5 | 305.41333 | 344.58 | 305.41333 | 369.82 | 305.41333 | 265.72 |
| $b \times 10^3$ | 0 | 0 | 9.4773 | 0.64172 | 0 | −1.7865 | 0 | −10.4158 | 0 | −14.8068 | 0 | −1.481 |
| $c \times 10^{-5}$ | 0 | −188.872 | −23.853 | 19.122 | −71.659732 | −212.58 | −71.659732 | 17.5334 | −71.659732 | 58.544 | −71.659732 | −208.72 |
| d | -4462 | −2039.4 | 271.930 | −448.48 | −1604.931 | 76.83 | −1604.931 | −2858.6 | −1604.931 | −3617.4 | −1604.931 | 100.92 |
| $e \times 10^{-8}$ | −12.605 | 23.1931 | 2.2801 | −3.7626 | 9.21837541 | 34.966 | 9.21837541 | -9.1142 | 9.21837541 | -15.7186 | 9.21837541 | 33.882 |
| $f \times 10^6$ | 0.000 | 0 | −1.1724 | −0.43907 | 0 | 0 | 0 | 0 | 0 | 0 | 0 | 0 |
| $g \times 10^9$ | 0 | 0 | 0.144915 | 0.048879 | 0 | 0 | 0 | 0 | 0 | 0 | 0 | 0 |
| $h \times 10^{-4}$ | 0 | 0 | 0 | −0.13326 | 0 | 0 | 0 | 0 | 0 | 0 | 0 | 0 |
| $V^0_{298.15}$ (cc/mol) | 113.16 | 125.38 | 11.250 | 16.76 | 66.2 | 62,66 | 66.2 | 60.674 | 66.2 | 62.51 | 66.2 | 64.84 |
| $A_{0,H}$ (kJ/mol) | 45.38 | – | 102.42 | – | 27.2478 | – | 28.155 | – | 25.304 | – | 25.304 | – |
| $A_{0,S}$ [J/(mol × K)] | 18.79 | – | 18.41 | – | 8.622 | – | 0 | – | 2.358 | – | 2.358 | – |
| $A_{0,V}$ (J/bar) | 0.1 | – | 0 | – | 0.0137 | – | 0.0137 | – | 0.0137 | – | 0.0137 | – |
| $A_{1,H}$ (kJ/mol) | 23.82 | – | 15.69 | – | 0 | – | 4.492 | – | 0 | – | 0 | – |
| $A_{1,S}$ [J/(mol × K)] | 0 | – | 0 | – | 0 | – | 0 | – | 0 | – | 0 | – |
| $A_{1,V}$ (J/bar) | 0 | – | 0 | – | -0.0071 | – | −0.0071 | – | −0.0071 | – | -0.0071 | – |
| $A_{2,H}$ (kJ/mol) | 0 | – | 0 | – | 0 | – | 0 | – | 0 | – | 0 | – |
| $A_{2,S}$ [J/(mol × K)] | 0 | – | 0 | – | 0 | – | 0 | – | 0 | – | 0 | – |
| $A_{2,V}$ (J/bar) | 0 | – | 0 | – | 0 | – | 0 | – | 0 | – | 0 | – |
| Mixing sites | 3 | | 1 | | 1 | | 1 | | 1 | | 1 | |

(continued)

**Table 4.4** (continued)

| Solid mixture | Ca-rich olivine | | Mg-rich olivine | | Melilites | | Wo-mix | | α-mix | | α′-mix | |
|---|---|---|---|---|---|---|---|---|---|---|---|---|
| End-member | $CaMgSiO_4$ | $Ca_2SiO_4$ | $Mg_2SiO_4$ | $CaMgSiO_4$ | $Ca_2Al_2SiO_7$ | $Ca_3MgSi_2O_7$ | $CaSiO_3$ | $MgSiO_3$ | $Ca_2SiO_4$ | $Mg_2SiO_4$ | $Ca_2SiO_4$ | $Mg_2SiO_4$ |
| $H^0_{f,298.15}$ (kJ/mol) | −2,255,027 | −2,318,090.3 | −2,173,000 | −2,255,027 | −3,988,158 | −3,862,527 | −1,632,094 | −1,496,564.4 | −2,291,172 | −2,089,329 | −2,303,787 | −2,089,329 |
| $S^0_{298.15}$ [J/(mol × K)] | 108.3 | 115.582 | 94.1 | 108.3 | 198.6 | 212.0 | 81.81 | 83.315 | 136.363 | 94.1 | 128.986 | 94.1 |
| a | 226.34 | 252.572 | 155.13 | 226.34 | 373.0874 | 387.396 | 149.07 | 133.75 | 252.572 | 155.13 | 252.572 | 155.13 |
| $b \times 10^3$ | 0 | 0 | 21.307 | 0 | 0 | 0 | 0 | −0.89325 | 0 | 21.307 | 0 | 21.307 |
| $c \times 10^{-5}$ | −11.797 | 0 | −44.060 | −11.797 | −47.78466 | 0 | −36.593 | −106.29 | 0 | −44.060 | 0 | −44.060 |
| d | −1542.7 | −2182.967 | 87.536 | −1542.7 | −2276.755 | −2938.8 | −690.3 | −38.415 | −2182.967 | 87.536 | −2182.967 | 87.536 |
| $e \times 10^{-8}$ | −0.2329 | 0 | 0 | −0.2329 | 4.779118 | −0.4079 | 4.8435 | −17.483 | 0 | 0 | 0 | 0 |
| $f \times 10^6$ | 0 | 0 | −0.7918 | 0 | 0 | 0 | 0 | 0 | 0 | −0.7918 | 0 | −0.7918 |
| $g \times 10^9$ | 0 | 0 | 0 | 0 | 0 | 0 | 0 | 0 | 0 | 0 | 0 | 0 |
| $h \times 10^{-4}$ | 0 | 0 | 0 | 0 | 0 | 0 | 0 | 0 | 0 | 0 | 0 | 0 |
| $V^0_{298.15}$ (cc/mol) | 51.48 | 57.5 | 43.67 | 51.48 | 90.08 | 92.52 | 39.62 | 31.33 | 53.6 | 43.67 | 51.8 | 43.67 |
| $A_{0,H}$ (kJ/mol) | 32.5 | – | 32.5 | – | 12 | – | 0 | – | −35.564 | – | −35.564 | – |
| $A_{0,S}$ [J/(mol × K)] | 0 | – | 0 | – | 0 | – | 0 | – | 0 | – | 0 | – |
| $A_{0,V}$ (J/bar) | 0.75 | – | 0.75 | – | 0 | – | 0 | – | 0 | – | 0 | – |
| $A_{1,H}$ (kJ/mol) | 0 | – | 0 | – | −12 | – | 0 | – | 0 | – | 0 | – |
| $A_{1,S}$ [J/(mol × K)] | 0 | – | 0 | – | 0 | – | 0 | – | 0 | – | 0 | – |
| $A_{1,V}$ (J/bar) | 0 | – | 0 | – | 0 | – | 0 | – | 0 | – | 0 | – |
| $A_{2,H}$ (kJ/mol) | 0 | – | 0 | – | 0 | – | 0 | – | 0 | – | 0 | – |
| $A_{2,S}$ [J/(mol × K)] | 0 | – | 0 | – | 0 | – | 0 | – | 0 | – | 0 | – |
| $A_{2,V}$ (J/bar) | 0 | – | 0 | – | 0 | – | 0 | – | 0 | – | 0 | – |
| Mixing sites | 1 | | 1 | | 1 | | 1 | | 2 | | 2 | |

### 4.5.2 $MgAl_2O_4$–γ-$Al_2O_3$

As to the mixing model adopted for spinel in this work, we recall that the alumina end member of the spinel solid mixture is conveniently described by using a Kröger-Vink notation (Kröger 1964) as $V_T^{||}(Al_{Al}^X Al_{Al}^X)_M(O_O^X O_O^X O_O^X V_O^{\cdot\cdot})_O$, where metal and oxygen vacancies are denoted as V; the dots and vertical bars at superscript indicate positive and negative excess charges, respectively; crosses at superscript indicate the neutral condition; subscripts identify the nominal site in the crystal lattice (T = tetrahedral, M = octahedral, O = oxygen site). This notation implies that vacancies in T and O sites are always coupled and thus have no configurational degrees of freedom. Assuming for simplicity the spinel component $MgAl_2O_4$ adopts a "normal" structure with no inversion between T and M sites at all temperature conditions, viz. $Mg_T^X(Al_{Al}^X Al_{Al}^X)_M(O_O^X O_O^X O_O^X O_O^X)_O$, the ideal mixing contribution reduces to the stoichiometric occupancy in the tetrahedral site.

Solution calorimetry of $MgAl_2O_4$–$Al_{8/3}O_4$ solid mixture carried out in a molten $2PbO{\cdot}B_2O_3$ solvent at 975 K by Navrotsky et al. (1986) indicated small negative heats of mixing "*relative to spinel standard states for both end-members*" which have been recast by the Authors in terms of a regular Redlich–Kister expansion with $A_{0,H} = -17.6$ kJ/mol. The $Al_{8/3}O_4$ compound utilized by Navrotsky et al. (1986) was actually constituted by commercial nanoparticles of cubic ($Fd\bar{3}m$) γ-$Al_2O_3$. Integrating the partial molar entropies of mixing of the pure aluminous component, based on the solubility data of Viertel and Seifert (1980) and Lejus (1964) and subtracting the ideal configurational contributions one obtains subregular entropic contributions sufficiently well reproduced by adopting $A_{0,S} = -25$ J/(mol K), $A_{1,S} = +5$ J/(mol K). In our (Kröger - Vink) notation, the purely aluminous spinel end member is $Al_2V_{Mg}^{||}O_3V_O^{\cdot\cdot}$, and we must simply account for the enthalpic and entropic standard-state contributions arising from structural deformation and defect energy formation. We may, to this purpose, adopt the procedure of Navrotsky et al. (1986) and use their solubility data scaled to the appropriate stoichiometry [×i.e., $\Delta H_{\alpha\rightarrow\gamma} = (3/4) \times 31.238 = 23.429$ kJ/mol, $\Delta S_{\alpha\rightarrow\gamma} = (3/4) \times 13.35 = 10.012$ J/(mol×K)]. With the $C_P = f(T)$ function assumed as unvaried, the resulting $H_f^o$ of γ-$Al_2O_3$ is enhanced to −1652.272 kJ/mol and S° is enhanced to 61.413 J/(mol×K), The Schottky-type defect state we imagined for our aluminous end member involves an increase of volume ($\Delta V_S$) due to the fact that the lattice around the vacant sites undergoes relaxation due to static effects. Because $\Delta V_S$ is inherently positive, the defect population (hence the expected solubility) is expected to decrease with increasing P. Indeed, based on the declared density of the commercial nanoparticles (2.8 g/cc), the molar volume of γ-$Al_2O_3$ is 3.641 J/bar. We adopt this value provisionally for γ-$Al_2O_3$ and assume no deviations from Vegard's Law. With the adopted parameterization, γ-$Al_2O_3$ at $T_f$ = 2348 K is 82 J/mol more stable than α-$Al_2O_3$ The equality of potentials is attained at T = 2340 K.

### 4.5.3 $MgO–Al_2O_3$

The solubility of alumina in MgO at subsolidus temperatures is quite limited and attains according to Alper et al. (1962) ~ 0.07 mol per mole of mixture at T = 2181 K. Based on this observation and on the experimental evidences of Whitney and Stubican (1971) on the conformation of the solvus limb at subsolidus conditions, the mixing properties are satisfactorily reproduces adopting a regular interaction parameter $A_{0,H}$ = 42.9 kJ/mol and enhancing $H_f^o$ of the $Al_2O_3$ end member in mixture (with respect to pure α-alumina) as indicated by Jung et al. (2004) (see Eqs. (4.39)–(4.40) and Table 4.4).

### 4.5.4 Sapphirines

Solid-state mixing between 442 and 351 polytypes may occur through a Tschermak-type coupled substitution, as more extensively discussed at para. 7.7.2. According to our computation, the MAS topology predicts P = 1 bar stability of the 351 polytipe at the expense of 442. Assuming ideal mixing on one site (consistent with coupled substitution), sappyrine mixtures are stable.

### 4.5.5 $MgSiO_3–AlAlO_3$ Orthopyroxenes

As noted by Ganguly and Ghose (1979), the Al/Mg substitution in Orthoenstatite is crystal-chemically limited to 25% in moles, thus leading to the "*orthopyrope*" (*Opy*) composition $Mg_3Al_2Si_3O_{12}$. According to Ganguly and Ghose (1979), the configurational entropy obeys the Al-avoidance principle, leading to a mixing contribution which attains, at this limiting composition, ~4 J/(mol K). The ensuing molar entropy of the limiting component $Mg_3Al_2Si_3O_{12}$ at T = 1300 K is 217.71 ± 1.36 J/(mol× K) (Ganguly and Saxena 1987). We adopt here for *Pbca* $AlAlO_3$ $K_0 = 250$ GPa, $K_1 = -50$ bar/K. $K_2 = 0$, $K' = 4.2$, $K'' = 0$; $\alpha_0 = 1.2805 \times 10^{-7}$, $\alpha_1 = 2.0303 \times 10^{-5}$. With this parameterization and the $C_P$ function of Ganguly and Saxena (1987), the isochoric heat capacity is approximately consistent with first principles up to 3000 K. To mimic the effect of Al-avoidance, we simply adopt a regular entropic interaction term $A_{0,S} = -10$ (Table 4.4).

### 4.5.6 Majorite-Pyrope

As already noted by Belmonte et al. (1017b) in the pressure range exceeding 17 GPa, majorite-pyrope solid mixture is the only (and most important) ternary liquidus phase in the MAS system.

By comparing the calculated HPA isotherms at P = 23 GPa in the T range 2973–3190 K with the results of high-pressure melting experiments carried out in the $Mg_4Si_4O_{12}$–$Mg_3Al_2Si_3O_{12}$ join (Kudo and Ito 1996; Gasparik 1992, 2003), it was observed that the liquidus temperatures predicted by HPa modeling for majorite-pyrope garnet are consistent with the observed ones within ±100 K, which is roughly the experimental uncertainty for this kind of measurements at HP-HT. We retain thus here the previous parameterization of the mixing properties, which assumed ideal mixing of the two end-members.

### 4.5.7 MgO–CaO

Based on the experimental observations of Doman et al. (1963) and Henney and Jones (1969), the miscibility gap at subsolidus is quite extended ($0.1 \leq XCaO \leq 0.8$ at $T \approx 2700$ K). The combined (visual + X-ray) observations allow us to depict the MgO–CaO solid mixture as subregular, with $A_{0,H} = 102.42$; $A_{0,S} = 18.41$; $A_{1,H} = 15.69$; $A_{1,S} = 0$ (Fig. 4.1; Table 4.4 and Eqs. (4.39)–(4.40)). We have no hints on eventual deviations from Vegard's Law, and we must assume provisionally ideality in this regard. The liquidus is satisfactorily conformed through a simple subregular enthalpic parameter ($A_{1,H} = -30$ kJ/mol). We will assume in the following that there is also a limited deviation from Vegard Law in the liquid state ($A_{0,V} = 0.2$ J/bar).

### 4.5.8 $Mg_2Si_2O_6$–$CaMgSi_2O_6$ Pyroxenes

The mixing properties of *C2c* pyroxenes have been investigated by many Authors and, although the differences among the estimated enthalpic interactions are minimal (Table 4.5), we may note a certain disagreement concerning the volumetric terms.

We may note also that the enthalpic interactions estimated through a Born parameterization of static potentials (Ottonello et al. 1992; Ottonello 1992) are in good agreement with the observations of Lindsley et al. (1981), Davidson and Lindsley (1989), therefore they will be adopted in the following.

For the other pyroxene polymorphs, we will adopt the parameterization of Jung et al. (2005) for enthalphy, entropy, and interactions. For volume interactions in $P2_1/c$, *Pbca*, and *Pbcn*, we will adopt the same parameters of *C2c* (Table 4.5).

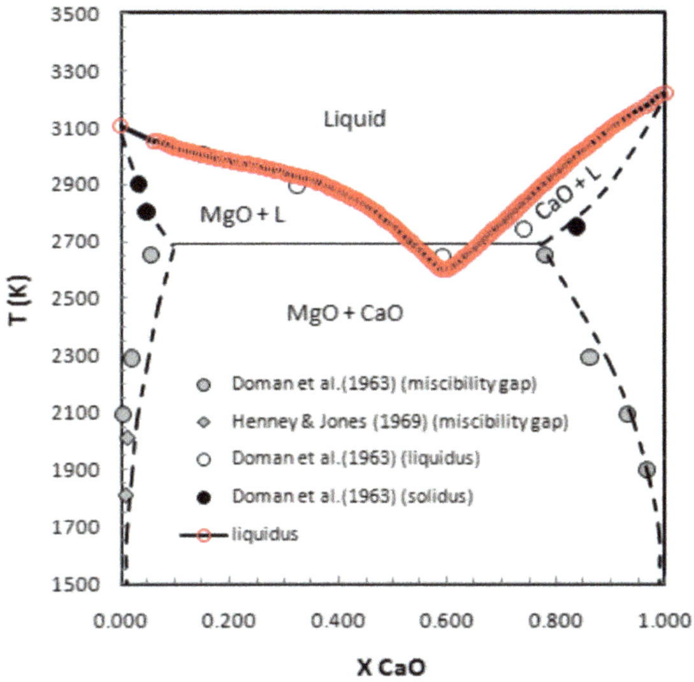


**Fig. 4.1** The MgO–CaO system at P = 1 bar. The red circles superimposed on the continuous liquidus curve are the discrete points obtained through HPA + convex-hull analysis in the ternary MCS system (see later)

**Table 4.5** Estimated enthalpic and volumetric interactions in *C2c* pyroxenes according to various authors

| $A_{0,H}$ | $A_{1,H}$ | $A_{0,V}$ | $A_{1,V}$ | Refs. |
|---|---|---|---|---|
| 21.905 | 0 | −0.0523 | 0 | (1) |
| 24.47 | 0 | 0.105 | 0 | (2) |
| 28.35 | 2.866 | 0.0101 | −0.0711 | (3) |
| 29.213 | 3.088 | −0.02255 | 0.01585 | (4) |
| 25.304 | 0 | – | – | (5) |
| 28.155 | 4.492 | 0.0137 | −0.0071 | (6) |

(1) Nickel and Brey (1984); (2) Holland et al. (1979); (3) Lindsley et al. (1981); (4) Davidson and Lindsley (1989); (5) Jung et al. (2005); (6) Ottonello et al. (1992), Ottonello (1992)

### *4.5.9 $Mg_3Al_2Si_3O_{12}$–$Ca_3Al_2Si_3O_{12}$ Garnets*

Figure 4.2 shows enthalpic interactions in binary aluminiferous garnet mixtures, according to various Authors, represented as sub-regular Margules models and

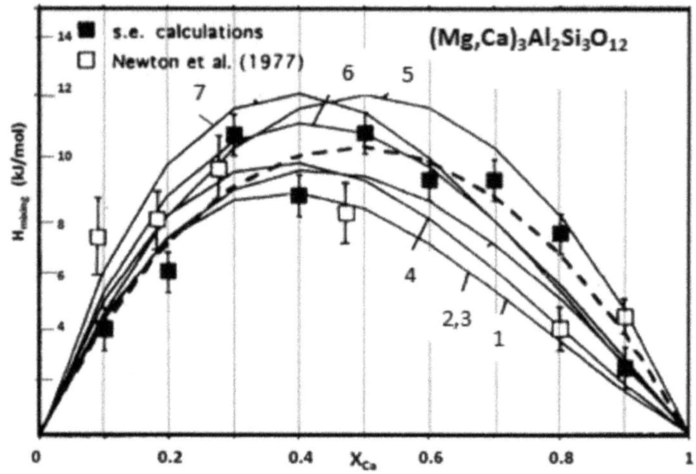


**Fig. 4.2** Enthalpic interactions in the (Mg, Ca)$_3$Al$_2$Si$_3$O$_{12}$ binary join.: experiments (▢); versus interionic potential calculations (■; Ottonello et al., unpublished), (1) Newton et al. (1977) calorimetric; (2) Ganguly and Saxena (1984) multivariate analysis; (3) Haselton and Newton (1980) calorimetric; (4) Wood (1988) phase equilibration; (5) Ganguly and Kennedy (1974) upper limit based on synthesis of pyrope—grossular solid mixtures and mechanical mixtures of end—members; (6) Ganguly et al. (1993) lattice strain theory; (7) Berman (1990) multivariate analysis

compared with calorimetric values and results of interionic potential model calculations. The overall picture delineates an almost symmetric interaction. The corresponding subregular Margules interaction parameters and their Van Laar counterparts are listed in Table 4.6. Since divalent cations (Mg, Ca) only occupy dodecahedral sites while octahedral sites are reserved for trivalent cations (Al), each cation has only one site typology at its disposal, and permutability is fixed by stoichiometry. Existing experimental data on the $Mg_3Al_2Si_3O_{12}$–$Ca_3Al_2Si_3O_{12}$ system show that the excess entropy of mixing terms is probably symmetric as well and coupled to a slightly positive excess volume (Ganguly et al. 1993).

## 4.5.10 $Mg_2SiO_4$–$Ca_2SiO_4$ Olivines

Figure 4.3 shows the Gibbs free energy of mixing generated at T = 873 K (P = 1 bar) by a Born-model parameterization of static potentials (Ottonello 1987). There are two solvi: one extending from $(Mg_{0.02}Ca_{0.98})_2SiO_4$ to $(Mg_{0.49}Ca_{051})_2SiO_4$, and one from $(Mg_{0.51}Ca_{0.49})_2SiO_4$ to pure forsterite. Monticellite is thus a stable compound with very limited solubility for $Ca_2SiO_4$ and practically no solubility for $Mg_2SiO_4$. The calculated excess energies (Table 4.4) are in good agreement with the mixing model of Warner and Luth (1973). The calculated enthalpy of formation at 298.15 K, 1 bar of *monticellite* (the stoichiometric compound $(Mg_{0.5}Ca_{05})_2SiO_4$) is −2258.945.

**Table 4.6** Estimated enthalpic (kJ/mol), entropic [J/(molxk)], and volumetric (J/bar) interactions in Mg-Ca aluminiferous garnet according to various authors. Margules (W) and Van Laar (A) parameters

| $W_{H,12}$ | $W_{H,21}$ | $W_{S,12}$ | $W_{S,21}$ | $W_{V,12}$ | $W_{V,21}$ | Refs. |
|---|---|---|---|---|---|---|
| 25.1 | 47.9 | | | | | (1) |
| 48.1 | 48.1 | | | | | (5) |
| 12.6 | 50.8 | 4.5 | 4.5 | | | (2) |
| 12.6 | 50.8 | | | | | (3) |
| 15 | 56.5 | | | | | (4) |
| 25.9 | 59.3 | | | | | (6) |
| 21.6 | 69.2 | | | | | (7) |
| 93.6 | 93.6 | | | | | (8) |
| 25.9 | 25.9 | 2.51 | 2.51 | 0.58 | 0.12 | (9) |
| 38.6 | 42.4 | | | | | (10) |
| | | 11.19 | 0 | 0.3 | 0 | (11) |
| | | | | 0.173 | 0.36 | (12) |
| $\mathbf{A_{0,H}}$ | $\mathbf{A_{1,H}}$ | $\mathbf{A_{0,S}}$ | $\mathbf{A_{1,S}}$ | $\mathbf{A_{0,V}}$ | $\mathbf{A_{1,V}}$ | **Refs.** |
| 36.5 | 11.4 | | | | | (1) |
| 48.1 | 0 | | | | | (5) |
| 31.7 | 19.1 | 4.5 | 0 | | | (2) |
| 31.7 | 19.1 | | | | | (3) |
| 35.75 | 20.75 | | | | | (4) |
| 42.6 | 16.7 | | | | | (6) |
| 45.4 | 23.8 | | | | | (7) |
| 93.6 | 0 | | | | | (8) |
| 25.9 | 0 | 2.51 | 0 | 0.35 | −0.23 | (9) |
| 40.5 | 1.9 | | | | | (10) |
| | | 5.595 | −5.595 | 0.15 | −0.15 | (11) |
| | | | | 0.2665 | 0.0935 | (12) |

(1) Newton et al. (1977); (2) Ganguly and Kennedy (1974); (3) Ganguly and Saxena (1984); (4) Haselton and Newton (1980); (5) Wood (1988); (6) Ganguly and Cheng (1994); (7) Berman (1990); (8) Hensen et al. (1975); (9) Ganguly (pers. Comm.); (10) Ottonello et al. (unpublished); (11) Newton et al. (1986); (12) Ganguly et al. (1993)

This value is not far from the values tabulated by Barin (1995) (−2261.870 kJ/mol) and Robie et al. (1978) (−2262.705), while the value assigned by Berman (1988) and here adopted is −2250.027 kJ/mol (Table 4.1).

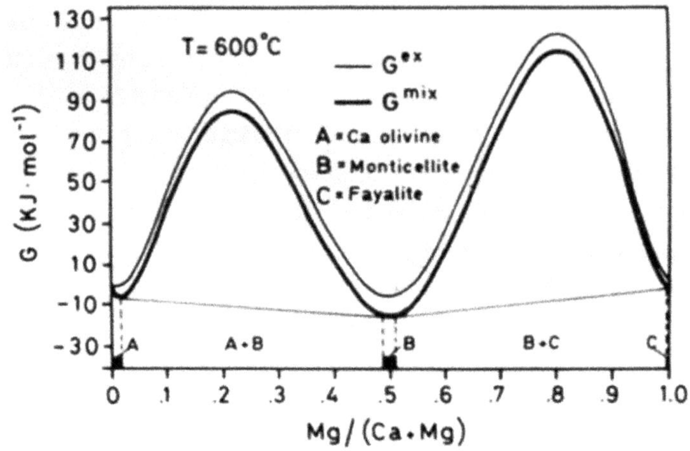


**Fig. 4.3** Gibbs free energy of mixing in the (Ca, Mg)$_2$SiO$_4$ join at T = 873 K. From Ottonello (1987) with modifications

### 4.5.11 $Ca_2Al_2SiO_7$–$Ca_2MgSi_2O_7$ Melilites

Gehlenite and Akermanite are the most important end members of *melilite* mixtures. Their mixing behavior has been calorimetrically investigated by Charlu et al. (1981), who observed a subregular excess of mixing here translated into a formal Van Laar expansion (see Eqs. (4.33)–(4.34) and Table 4.4). It is well known that, in terms of volume, the mixture deviates in a complex way from ideality and undergoes phase transitions (Swainson et al. 1992; Merlini et al. 2005; Merlini et al. 2009a, b). We disregard here, however, these additional complexities.

### 4.5.12 $CaSiO_3$–$MgSiO_3$ (Wollastonite Type)

We assume for the Mg substitution of Ca in wollastonite, random mixing on one site with no deviation from Vegard's Law, and absence of excess enthalpy.

### 4.5.13 A (Larnite-Type) α′ bredrigite-type $Ca_2SiO_4$–$Mg_2SiO_4$ Mixtures

Jung et al. (2004) assume for these mixtures random mixing over two sites and a regular excess enthalpy $A_{0,H} = -35.564$ (kJ/mol). We adopt here their assessment and assume no deviation from Vegard's Law.

## 4.6 The Liquid

In the Hybrid Polymeric Approach (HPA), the Gibbs free energy of mixing of multi-component polymeric substances is seen as a sum of discrete contributions arising from different forms of interaction, namely a dominant chemical interaction taking place among polymeric units (structons), a (subordinated) interaction taking place in the cation matrix, and a strain energy contribution arising from medium range disorder of the various structons within the anion matrix. Different from the previous model releases, here we treat the effect of deviation from Vegard's Law ($PV_{excess}$):

$$G_{mixing} = G_{chemical} + G_{cat.mix\text{ - }id} + G_{cat.mix\text{ - }exc} + G_{strain} + G_{polar} + PV_{excess} \quad (4.35)$$

and treat the high-order interaction terms present in some second-rank and third-rank simplexes as charge polarization effects ($G_{polar}$). In the first four joins of interest in this study, there is only a single Lux-Flood basic oxide (MgO or CaO) interacting with a single acidic oxide ($SiO_2$ or $Al_2O_3$), hence the second and third terms in the above expansion are absent. Also, the fifth join may be treated in the same manner, for computational simplicity. MgO and CaO are Network Modifiers (NM) while $SiO_2$ and $Al_2O_3$ are Network Formers (NF). The chemical interaction between network formers and network modifiers constitutes by far the main form of energy gained by the liquid in the de-polymerization process[1]:

$$G_{chemical} = \frac{(O^-)}{2} RT \ln K_P \quad (4.36)$$

In the above equation, $(O^-)/2$ is the number of contacts among functionals of the polymeric units of the Toop-Samis model (Toop and Samis 1962a, b), and $K_P$ is the polymerization constant that describes the gain in chemical energy acquired by the melt in a single interaction among the various kinds of oxygen:

$$K_P = \frac{(O^{2-}) \times (O^0)}{(O^-)^2} \quad (4.37)$$

$$G_{chemical} = \frac{(O^-)}{2} RT \ln K_P \quad (4.38)$$

In the above equation, $(O^-)/2$ is the number of contacts among functionals of the polymeric units of the Toop-Samis model (Toop and Samis 1962a, b), and $K_P$ is the polymerization constant that describes the gain in chemical energy acquired by the melt in a single interaction among the various kinds of oxygen:

[1] It may be easily shown that the reaction between a NM oxide (e.g MgO) and a NF ($SiO_2$) (hence a Gibbs free energy of mixing term to be added to the energy of pure components) is a de-polymerization process (Ottonello, 2024a,b).

$$K_P = \frac{(O^{2-}) \times (O^0)}{(O^-)^2} \tag{4.39}$$

The molar amount of $(O^-)$ is one of the two eigenvalues of the quadratic equation in $(O^-)$ solved for discrete molar amounts of oxygen acceptors (NF) and $K_P$ (Toop and Samis 1962a, b). However, at intermediate-low acidity (i.e. tipically NF $\leq 0.55$) the amount of free oxygen ions in the melt computed via mass balance in the assumption of a completely basic behavior of oxygen donors (NM) is too low with respect to a Temkin-like activity of the oxide and a maximum depolymerization limit (monomer) must be introduced along the guidelines anticipated long ago for the CaO-FeO-$SiO_2$ slags (Ottonello 2001). The polymerization constant is intimately linked to the donor pressure of the system and depends on the contrast in the Lux-Flood acid–base characters of oxygen donors NM and oxygen acceptors NF. In an NM-NF system of high contrasting acid–base character, such as MgO-$SiO_2$, the chemical interaction is mainly enthalpic in nature: i.e., with $\ln K_P$ expressed as a function of T ($\ln K_P = A/T + B$), the A term is dominant and the B term limited.

The chemical interactions within the anion matrix of an *n*-component system are completely fixed by the interaction properties observed along the $n - 1$ limiting binaries (Moretti and Ottonello 2003). The intrinsic thermodynamic significance of this evidence may be reduced to the Flood-Grjotheim thermodynamic cycle (Flood and Grjotheim 1952). Denoting $K_{P,A-F}$, $K_{P,B-F}$ the polymerization constants valid along the limiting A–F B–F binaries, the polymerization constant of the extended Toop-Samis model, in fact, turns out to be:

$$\ln K_P = N'_{A^{v+}} + \ln K_{P,A-F} + N'_{B^{0+}} + \ln K_{P,B-F} + \ldots \tag{4.40}$$

The above equation states that in a compositionally complex melt network, modifier cations A, B of charge $v_A^+$, $v_B^+$ interact with free oxygen $O^{2-}$ (and hence with the polymeric units built up by the network former F) with a simple proportionality described in terms of $N'_{A^{v+}}$, $N'_{B^{v+}}$ "*electrically equivalent fractions*" (Ottonello 2001):

$$N'_{A^{v+}} = \frac{v_A^+ n_A}{v_A^+ n_A + v_B^+ n_B + \ldots} \tag{4.41}$$

It is thus possible to depict the thermodynamic properties of an *n*-component system through Toop's asymmetric deconvolution (i.e., with the assumption that the cation matrix is not appreciably affected by the anions, and the anion matrix is affected by cations).

The Hookian strain contribution, interpreted as the energy spent (or gained) in conforming the relative arrangement of the various polymers in the anion sublattice (Ottonello 2005), has been recently shown to be of a vibrational nature (Ottonello 2024a, b).

There is a general consensus on the fact that this form of energy is largely entropic. In copolymer blends, such as rubberlike materials, this energy is associated with

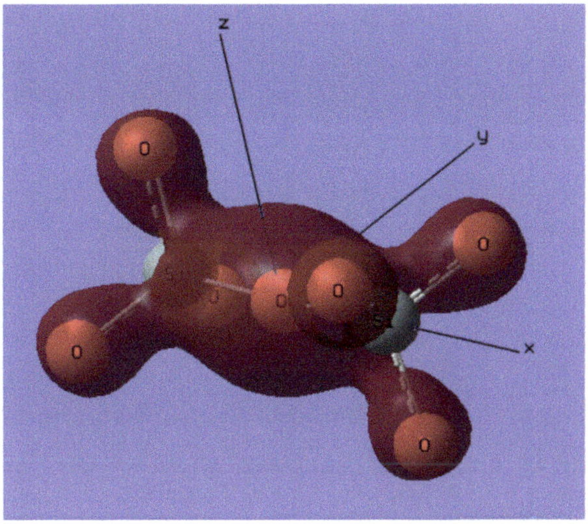


**Fig. 4.4** The σ-bonding orbitals of the $[Si_2O_7]^{6-}$ dimer resolved at B3LYP/TZVP theory level. The isodensity limit is 0.03 e/Å$^3$. The electron cloud encompasses all oxygen centers in the dimer. In analogy to what is observed in silicate and silico-aluminate rings, this MO is interpreted as a Pauling's double bond (Ottonello et al. 2010a; Ottonello 2024a, b)

conformational adjustments of polymer chains (see Fredrickson et al. 1994, Bates and Fredrickson 1999; and references therein for an extended discussion). In silicate liquids, this amount of entropy is associated with a limited increase in the vibrational frequency of Rigid Unit Modes (RUM) affecting adjacent monomeric groups in polymer chains and induced by the presence of NM cations coulombically contouring the structures. As such, entropy does not act as a "restoring" effect, such as in rubberlike materials, but rather favors decomposition. As an example, we see in Fig. 4.4 the structure of the simple dimeric chain $[Si_2O_7]^{6-}$ in a continuum with a permittivity $\varepsilon_0 = 3.8$ resolved at DFT/B3LYP level by applying Tomasi's Polarized Continuum Model (Ottonello et al. 2010a).

The first three optical modes of this linear dimer are RUMs, and the second one, with $v_2 = 75.265\ cm^{-1}$ is a bending motion (Fig. 4.5). Treating this RUM as a simple Einstein's oscillator, we may obtain its contribution to the vibrational entropy of the liquid by applying:

$$S_{v_1} = R\left[\frac{X_1}{e^{X_1} - 1} - \ln\left(1 - e^{X_1}\right)\right] \tag{4.42}$$

where $X = \frac{2\pi v_L h}{k}$, with $v_L$ = speed of light and h, k respectively the Planck's and Boltzmann constants, is the a-dimensional frequency. A simple increase of 10 $cm^{-1}$ in vibrational frequency causes a decrease of −0.52 J/[mol K] virtually unaffected by T in the T-range of interest (i.e., from 1000 to 3000 K).

Because there is a direct relation between linear frequency v and molecular force constant; i.e.

$$\Delta F_K = 4\pi^2 m^* \times \left[(\nu + \Delta\nu)^2 - \nu^2\right] \tag{4.43}$$

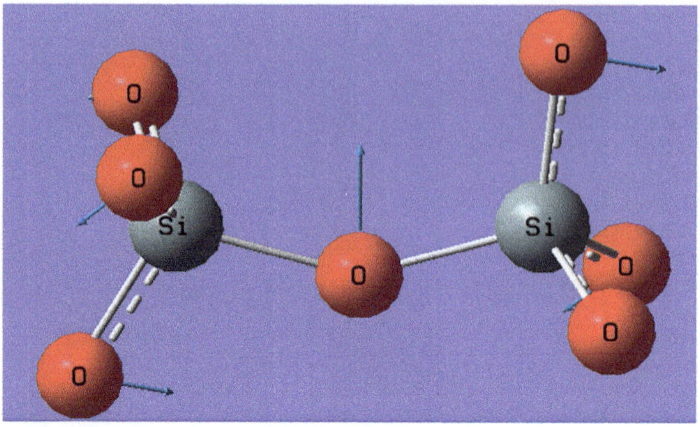


**Fig. 4.5** Bending Rigid Unit Mode (RUM) of the $[Si_2O_7]^{6-}$ dimeric unit in a Polarized Continuum with $\varepsilon_0 = 3.8$ (Ottonello et al. 2010a). The force constant of the RUM is 0.0534 mDyne/Å and the reduced mass m* is 16.0034 AMU. Displacement vectors are superimposed to visualize the kind of motion. From Ottonello (2024a; with modifications)

with m* = reduced mass of the molecule, this increase of vibrational frequency may be ascribed to a slight increase in the force constant of the bond $F_K$ (~0.015 mDyne/A) caused by the stabilizing effect that NM cations have on the low-lying σ-bonding Molecular Orbitals (MOs) of the oxygen framework, such as the Pauling's double bond of oxygen acting on next-neighbor oxygens in silica and silico-aluminate rings (Gatti et al. 2012), and, most likely, on linear chains as well (Richet and Ottonello 2014).

The vibrational contributions are still counted in terms of contacts

$$\Delta G_{strain} = \frac{(O^-)}{2}\eta_{strain} \tag{4.44}$$

And the η parameter is conveniently expanded as a function of the molar fraction of Network Formers in the system ($X_{NF}$) and T:

$$\eta_{0,strain} = \eta_{0,H} - T \times \eta_{0,S} \tag{4.45}$$

$$\eta_{strain} = \sum_{k=0}^{n} \eta_{k,\ strain} \cdot (2X_{NF} - 1)^n \tag{4.46}$$

The non-ideality of the silicate liquids in terms of volumes cannot be easily recast into first principles. Deviations from Vegard's Law may, however, be accounted for by a Van Laar model based on a Redlich–Kister expansion:

$$V_{exc} = \sum_{k=0}^{n} \eta_{k,V} \cdot (2X_{NF} - 1)^k \cdot [X_{NF} \cdot (1 - X_{NF})] \tag{4.47}$$

This notation (4.47) stresses that the P-dependent excess volumes are constrained by experimental observations at discrete P conditions and are not based on a Toop-Samis rationale involving the relative amount of ($O^-$) quasi-species.

Disregarding this last form of energy, the advantage of polymeric models with respect to more agnostic procedures is that the chemical interaction assumes a precise significance in terms of energy, based on the Lux-Flood acid–base reaction formalism. The number of contacts among polymer functional (corresponding to half the molar amount of singly bonded oxygen) is in fact quite similar to the number of contacts of the Quasi-chemical model (Guggenheim 1935; Fowler and Guggenheim 1939) but it depends explicitly on the contrast in the donor and acceptor properties of the basic and acidic oxides in mixture, expressed in terms of Pauling's electronegativity differences or optical basicity contrast (Ottonello 2005; and references therein). When the two oxides differ greatly in their acid–base properties, then the interaction is expected to be almost purely enthalpic and not significantly affected by temperature. As we will see later on, these assessments are strengthened by indirect estimates of the enthalpy of mixing per unit mole of melt, which may be obtained from experimental heats of fusion of solids nucleating in the system.

In the upper part of Table 4.7 are listed the HPA parameters for the first five joins of the system. In the $Al_2O_3$–$SiO_2$ join, $Al_2O_3$ is treated as an NM and $SiO_2$ as an NF. In the $6^{th}$ join (i.e., CaO-MgO; not shown in Table 4.7), we have only a cationic interaction, because the anion matrix is composed entirely of free $O^{2-}$ ions.

In the second-rank simplexes CAS, MAS, CMA, CMS, which limit the tetrahedron, besides the various forms of energy previously discussed, one observes the presence of ternary interactions which may be treated agnostically, such as in MQC, or ascribed to physical processes as in HPA.

The simplest numerical procedure to account for ternary interactions is the symmetrization of the compositional space in order to observe the highest interaction at the center of the barycentric coordinates and zero contributions at the binary limits. The molar proportions of components are thus altered by multiplication by stoichiometric coefficients in order to obtain appropriate equivalent fractions. This is practically the expedient fostered by Pelton and Blander (1986) in the presentation of the MQC approach.

It would, however, be nice to ascribe to these coefficients a first-principles significance. It is rather evident that, because calculations indicate that the interaction is merely enthalpic, it cannot be ascribed to it any purely configurational attribute. We recall to this purpose that, according to Maxwell's equations, when a charge *e* (taken as origin) is placed in a dielectric medium consisting of discrete ions, the polarization work $W_P$ is:

**Table 4.7** Parameters of the hybrid polymeric approach for the CMAS system along binary joins. $\ln K_P = A + B/T$. $\eta_{k,V} = \sum_{m=-1}^{4} \eta_{k,m,V} \cdot P^m$

| Join | $MgO–SiO_2$ | $CaO–SiO_2$ | $MgO–Al_2O_3$ | $CaO–Al_2O_3$ | $Al_2O_3–SiO_2$ |
|---|---|---|---|---|---|
| A | −3330 | −14,416 | −3350 | −10,000 | 0 |
| B | −2.5 | 0 | −1.74 | −1.6 | −1.85 |
| $\eta_{1,S}$ | 3[a] | −2.5[a] | 0 | −26 | 0 |
| $\eta_{2,S}$ | 0 | 0 | 0 | 32 | 0 |
| $\eta_{2,H}$ | 0 | 0 | −12 | 0 | 0 |
| $\eta_{3,S}$ | 0 | 0 | 0 | 16 | 0 |
| $\eta_{4,S}$ | 0 | 0 | 0 | 11 | 0 |
| $\eta_{5,S}$ | 0 | 8[a] | 0 | 0 | 20[a] |
| $\eta_{6,S}$ | −10[a] | 0 | 0 | 0 | 13[a] |
| **P (GPa)** | **≤ 30** | **≤ 15.2** | | | |
| **k** | **0** | **0** | | | |
| $\eta_{k,,0,V}$ | −2.282E-1 | −6.235E-1 | – | – | – |
| $\eta_{k,1,V}$ | 6.273E-3 | 7.728E-2 | – | – | – |
| $\eta_{k,2,V}$ | −1.255E-4 | −2.433E-3 | – | – | – |
| **k** | **1** | **5** | | | |
| $\eta_{k,0,V}$ | −1.390E-1 | 1.17 | – | – | – |
| $\eta_{k,1,V}$ | 5.184E-2 | −1.823E-1 | – | – | – |
| $\eta_{k,2,V}$ | −9.006E-4 | 7.559E-3 | – | – | – |
| **P (GPa)** | **≥ 30** | **≥ 15.2** | | | |
| **k** | **0** | **0** | | | |
| $\eta_{k,0,V}$ | −2.510E-1 | −0.02 | – | – | – |
| $\eta_{k,1,V}$ | 6.878E-3 | – | – | – | – |
| $\eta_{k,2,V}$ | −3.992E-5 | – | – | – | – |
| **k** | **1** | **5** | | | |
| $\eta_{k,,0,V}$ | 1.137E + 0 | 0.15 | – | – | – |
| $\eta_{k,1,V}$ | −2.0835E-2 | – | – | – | – |
| $\eta_{k,2,V}$ | 9.169E-5 | – | – | – | – |

[a] limited to the acidic part of the join, i.e., $X_{NF} \geq 0.5$

$$W_P = -\frac{1}{2} e\phi = -\frac{1}{2} e \sum_i \frac{\alpha_i F_i}{r_i^2} \tag{4.48}$$

where $\phi$ is the potential field, $\alpha_i$ are ionic polarizabilities of the discrete ions, $r_i$ are their distances from the origin, and $F_i$ is the component of the total electric field directed toward the central charge. Assuming for simplicity the ionic polarizabilities and interionic distances in Eq. (4.48) to vary linearly with composition, the excess polarization energy ($\Delta G_{excess}$) may be recast in terms of the total electric field components, split into partial charges ($\Delta$) and a scalar term (Q). Assuming that each cation contributes with a discrete and constant charge fraction $\delta_i$ to polarization we may pose:

$$\Delta = \frac{\prod_{i=1}^{n} X_i \delta_i}{\left(\sum_{i=1}^{n} X_i \delta_i\right)^n} \tag{4.49}$$

Since the Faraday constant ($F$) is related to the elementary charge $e$ through the Avogadro number $N_0$ (i.e., $e = F/N_0$), we may further pose:

$$\Delta G_{polar} = \Delta W_P = -\frac{1}{2} Q \times \Delta \times F \tag{4.50}$$

where Q is the scalar term returning the correct value of the interaction when expressed in Volts ($F = 96.485$ kJ/V). Charge polarization is effective only whenever some charge transfer between NM and NF occurs to keep charge homogeneity in the anion matrix. Nonlinear inverse procedures applied to the MAS and CAS simplexes suggest that polarization effects inducing apparent symmetry in the ternaries are quite simple. In the MAS system, we get $\delta_{Mg} = 1.0$, $\delta_{Al} = \delta_{Si} = 0.5$ with $\delta_{Mg}$ decreasing progressively from 1.0 to 0.5 when the molar amount of MgO exceeds $Al_2O_3$ (Belmonte et al. 2017b) and $Q = -4.871$ V. In the CAS system we get $\delta_{Ca} =$ 1.0 (constant), $\delta_{Al} = 0.5$, $\delta_{Si} = 0.3$; $Q = -6.508$ V. In the CMA and CMS simplexes, there is no charge compensation, and in the third-order simplex CMAS, we may safely assume that the polarization effects proceed linearly with the Mg(Ca + Mg) ratio. The amount of polarization energy is in all cases quite subordinate with respect to all the other forms of energy (see later on).

# Chapter 5
# End-Member Components

The correctness of thermodynamic and thermophysical parameters of pure components is obviously crucial to obtain sound results, and the obedience to first principles is mandatory. The adoption of quantum–mechanical procedures is of great help in this sense.

## 5.1 $SiO_2$

### *5.1.1 $SiO_2$ Liquid*

An *ab-initio* investigation of the structure-energy and vibrational properties of silica clusters in a dielectric continuum through hybrid DFT gas-phase calculations coupled with a Polarized Continuum Model approach has shown that an aggregate of $D_{6h}$ network $SiO_2$ units and $[SiO_4]^{4-}$ monomers locally ordered in the short-medium range reproduces well both the experimentally observed isobaric heat capacity of silica at low-T and the deviation from the Debye $T^3$ law (Ottonello et al., 2010a). Furthermore, the rotational and translational components generate a heat capacity gap at the melting temperature ($\Delta C_P$), which is only partly counterbalanced by the loss of coherent motion of all the atoms in the ensemble. The thermophysical parameters of liquid silica were optimized by Belmonte et al. (2017a) in order to be consistent with the observed melting curves of the different $SiO_2$ polymorphs in the low- to high-pressure range. The guidelines of this assessment can be found elsewhere (Ottonello et al., 2013). The melting curve of $SiO_2$ polymorphs was constrained according to those observed by Jackson (1976) for cristobalite and β-quartz, Dalton and Presnall (1997) for coesite, and Shen and Lazor (1995) for stishovite. The adoption of *ab-initio* properties for stishovite (Ottonello et al., 2009a) led to an improvement of the $SiO_2$ melting curve at HP-HT with respect to our previous investigation (Ottonello et al., 2013). Belmonte et al. (2017a) refined the thermophysical properties of the

© The Author(s), under exclusive license to Springer Nature Switzerland AG 2026 
G. A. Ottonello et al., *Ab-initio Assisted Petrology*,
https://doi.org/10.1007/978-3-032-25658-4_5

liquid silica on the basis of a new *ab-initio*-assisted assessment of the thermodynamic properties of $\alpha$-cristobalite, coupled with the effects of the $\alpha/\beta$ lambda transition. The molar volume of the fictive liquid at standard state was lowered from 2.782 to 2.6 J/bar, and the baric derivative of the bulk modulus was set to = 12.8 (instead of 15.8) (Table 5.2). All the remaining thermodynamic and thermophysical parameters necessary to depict the properties of the liquid (Tables 5.1 and 5.2) are the same as Belmonte et al. (2017a). We may see in Fig. 5.1 the (*ab-initio* assisted) PVT behavior of the $SiO_2$ liquid, in a wide P–T range, compared with two limiting isotherms of the AIMD investigation of Karki et al. (2007).

### 5.1.2 Silica Polymorphs

The assessment of the thermophysical parameters for all $SiO_2$ polymorphs was carried out by Belmonte et al. (2017a) and, with scanty or contradictory experiments, was based on first-principles (Demuth et al., 1999; Ottonello et al., 2009a, 2010a). Because the rheology of a substance is quite important in establishing its extrinsic stability limits at the high thermo-baric regimes ruling in the planetary interiors, the properties of the most dense crystalline polymorph (Stishovite) assumed in this assessment a key role and will be outlined again as first. The adopted thermochemical parameters for most of the silica polymorphs are those of Berman (1988). This choice is dictated by the fact that Berman (1988) adopted $\alpha$-Quartz as a reference phase for the enthalpy of formation from the elements for a large number of solids, and his model-depicted extrinsic stability limits for the low-density polymorphs of silica are still unmatched in their accuracy with respect to experiments. Finally, as we will see later on, the entropy of $\alpha$ and $\beta$ forms of Cristobalite of Berman (1988) are in full agreement with first principles.

#### 5.1.2.1 Stishovite

Stishovite, the rutile-type $P4_2/mnm$ tetragonal aggregation form of silica, has been investigated by us *ab-initio* at the DFT/B3LYP theory level under the General Gradient Approximation (GGA) (Ottonello et al., 2009a). It is well known that the procedure results in a slight overestimation of the molar volume at the athermal limit, so we adopted in this study the lower limit in experimental observations, which range from 1.4000 to 1.4036 J/bar (Bassett and Barnett, 1970; Ito et al., 1974; Sugiyama et al., 1987; Ross et al., 1990; Li et al., 1996; Brazhkin et al., 2005). The quasi-harmonic gamma-mode analysis of the substance was later pushed to 5000 K; a value more consistent with the high-T regimes that may be attained in the planetary interior (Belmonte et al., 2017a; see Eqs. 4.3–4.8).

The analysis confirmed that $\alpha_T K_T = f(T)$ has at high T a constant slope attaining 0.0059 GPa/K at T = 5000 K. Our previous $\alpha_T = f(T)$ (Ottonello et al. 2009a) was then remodeled, constraining the thermal derivative dK/dT to achieve consistency over the

**Table 5.1** Thermodynamic properties of the pure liquid components in the CMAS system. $H^0_{f,298}$ = enthalpy of formation from the elements at $T_r = 298.15$ K, $P_r = 10^5$ Pa; $S^0_{298}$ = standard-state entropy of the substance; $C_P$ = isobaric heat capacity ($C_p = \mathbf{a} + \mathbf{b}T + \mathbf{c}T^{-2} + \mathbf{d}T^{-1/2} + \mathbf{e}T^{-3} + \mathbf{f}T^2 + \mathbf{g}T^3 + \mathbf{h}T^{-1}$)

| Component | $H°_f$ | S° | a | b x $10^3$ | c x $10^{-5}$ | d | e x $10^{-8}$ | f x $10^6$ | g x $10^9$ | h x $10^{-4}$ | Notes |
|---|---|---|---|---|---|---|---|---|---|---|---|
| | (J/mol) | [J/(mol × K)] | | | | | | | | | |
| $SiO_{2,\ liquid}$ | −911,746.2 | 33.8870 | *86.564* | *−0.42937* | *−48.527* | *−114.33* | *7.2829* | *0.02502* | *0.00014279* | – | (1) |
| | | | 1.891 | −2.572 | – | – | – | 0.68830 | 0.00449600 | – | |
| | | | **88.455** | **−3.00137** | **−48.527** | **−114.33** | **7.2829** | **0.71332** | **0.00592390** | **–** | |
| $Al_2O_{3,\ liquid}$ | −1,640,114.8 | 6.3630 | **165.09** | **2.3524** | **−80.211** | **396.37** | **9.5228** | **−0.1303** | **0.00000000** | **–** | (2) |
| $MgO_{liquid}$ | −563,716.5 | 16.7250 | *72.035* | *−3.0187* | *−13.124* | *−257.27* | *3.0899* | *0.37792* | *−0.015304* | – | (3) |
| | | | 0.673 | 4.8370 | – | – | – | 0.00115 | −0.0227800 | – | |
| | | | **72.708** | **1.8183** | **−13.124** | **−257.27** | **3.0899** | **0.37907** | **−0.0380840** | **–** | |
| $CaO_{\ liquid}$ | −569,918.0 | 56.0067 | **66.181** | **3.4509** | **14.209** | **−561.38** | **−3.6202** | **0.38087** | **−0.1024700** | **0.2681** | (4) |

(1) Belmonte et al. (2017a): Isobaric heat capacity equation (values in bulk) obtained by adding anharmonicity terms to the DFT-PCM vibrational terms of D6h network units and $[SiO4]^{4-}$ monomers in a medium of permittivity $\varepsilon = 3.8$ [values in italics, Ottonello et al. (2010)]; Standard state enthalpy of formation from the elements and standard state entropy based on thermodynamic cycle calculations (see text).

(2) This work. Based on the DFT-PCM calculations of Belmonte et al (2013: *D3d6* network units and $[AlO4]^{5-}$ monomers in a medium of permittivity $\varepsilon = 4.575$. Heat capacity function refitted and valid up to 5000 K.

(3) Belmonte et al. (2016): anharmonicity terms added to the DFT-PCM vibrational terms of $[MgO4]^{6-}$ monomers $[Mg]^{2+}$ cations in a medium of permittivity $\varepsilon = 13.3$ and vibrational and rotational contributions.

(4) Belmonte et al. (2017a). See text

**Table 5.2** Thermoelastic properties of the liquid components assessed on the melting curves of the solid polymorphs ($\beta$ -cristobalite, $\beta$ -quartz, coesite, and stishovite for $SiO_2$; periclase for MgO; corundum for $Al_2O_3$; lime for CaO). $V^0{}_{298}$ = standard-state molar volume ($T_r = 298.15$ K, $P_r = 10^5$ Pa); $K_0$ = bulk modulus at the athermal limit ($T = 0$ K and $P = 0$ GPa); $\alpha_V$ = volume thermal expansivity ($\alpha_V = \alpha_0 T + \alpha_1 + \alpha_2 T^{-1}$). $K_0$ and $K''_{(P,,0)}$ of liquid MgO are from Karki et al. (2006). References as in Table 5.1

| Component | $V^0{}_{298.15}$ (cc/mol) | $K_0$ (GPa) | $K'$ | $K_1$ (bar/K) | $K_2$ (bar/$K^2$) | $K''$ ($K^{-1}$) | $\alpha_0 \times 10^7$ | $\alpha_1 \times 10^7$ ($K^{-1}$) | $\alpha_2 \times 10^3$ ($K^{-2}$) | References |
|---|---|---|---|---|---|---|---|---|---|---|
| $SiO_{2,\ liquid}$ | 26.0 | 5.0 | 12.8 | 0 | 0 | 0 | −0.3611 | 843.1 | 0 | [4] |
| $Al_2O_{3,\ liquid}$ | 37.392 | 5.485 | 19.5 | 0 | 0 | 0 | −0.2145 | 548.2 | 0 | [4] |
| $MgO_{liquid}$ | 11.41 | 35.3 | 4.81 | 0 | 0 | 0 | −0.05514 | 1088.0 | 8.412 | [3] |
| $CaO_{\ liquid}$ | 16.2 | 36.0 | 4.5 | −45 | 0 | 0 | 0 | 1110.0 | 0 | [4] |

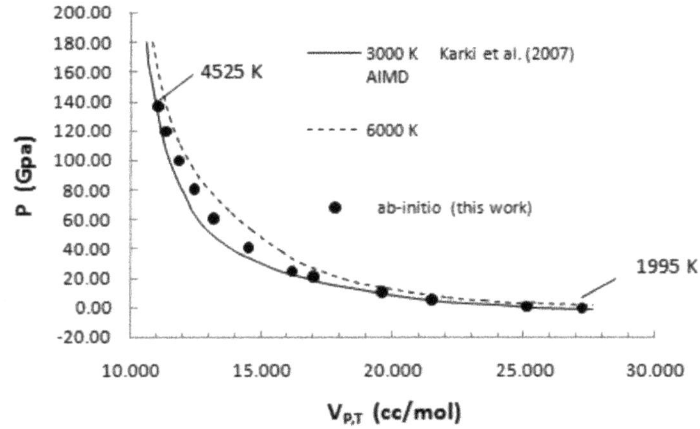


**Fig. 5.1** Volumes of liquid $SiO_2$ sampled at discrete points along the melting curve and compared with the AIMD results of Karki et al. (2007). At the mantle/core boundary, we obtain a melting temperature of 4525 K, assuming stishovite to be the stable polymorph

entire T range of interest. The results of this exercise are summarized in Table 4.2. The bulk modulus at the athermal limit obtained in this way ($K_0 = 270$ GPa) is somewhat higher than the value obtained from the simple static analysis of the potential well ($K_0 = 250$ GPa) and intermediate between the values derived respectively from General Gradient Approximation (GGA) and Local Density Approximation (LDA) analyses (see Demuth et al.,1999; and references therein). Because the anharmonicity at high T is slightly affected by the introduced modifications, Shomate's parameters of the isobaric heat capacity were re-fitted by non-linear minimization procedures (Table 4. 1). The compressional behavior of Stishovite in a wide baric range consistent with the pressure attained in the planetary interior is shown in Fig. 5.2 Experimental observations in the low-intermediate P range are superimposed for comparative purposes.

In Fig. 5.3a are shown the $P = 10^5$ Pa *ab-initio* isochoric ($C_V$) and isobaric ($C_P$) heat capacities of Stishovite, compared with low-T $C_P$ experimental observations. The *ab-initio* computation is in good agreement with the low-T measurements of Holm et al. (1967) as well as with the calorimetric experiments of Akaogi et al. (1995) at intermediate T conditions.

The isochoric heat capacity $C_V$ obviously obeys the Dulong-Petit rule at high temperature, and the isobaric heat capacity $C_P$, while very consistent with low-T experiments, converges toward the $3nR + TV\alpha^2K$ condition dictated by anharmonicity. The anharmonicity is definitely non-negligible and attains ~14.7 J/(mol × K) at the maximum T of investigation. The adopted entropy of the substance at standard state [$S° = 27.809$ J/(mol × K)] is that one obtained *ab-initio*, (Ottonello et al., 2009a) while the enthalpy of formation from the elements at standard state of Stishovite was selected in order to reproduce the invariant point of coexistence with liquid and Coesite and the melting curve of the substance (see later). It is somewhat

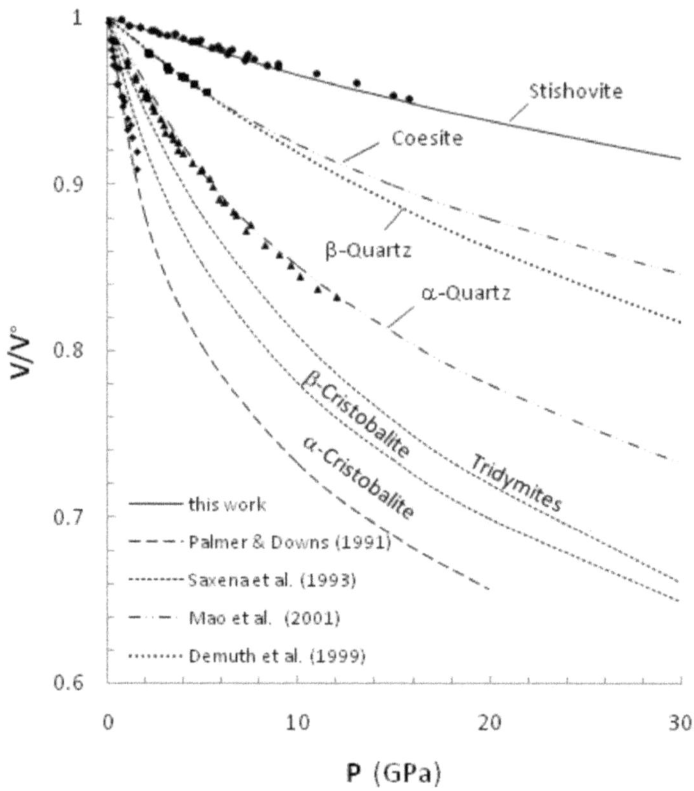


**Fig. 5.2** Compressional behavior of Silica polymorphs, experiments vs assessments. Stishovite: Olinger (1976); Ross et al. (1990); Liu et al. (1999). Coesite: Levien and Prewitt (1981); α-Quartz: Vaidya (1973); Olinger and Halleck (1976); D'Amour et et al. (1979); Levien et al. (1980); Cartz and Jorgensen (1981); Jorgensen (1978). α-Cristobalite: Palmer and Downs (1991). From Belmonte et al., 2017a, with modifications

lower than current estimates (i.e. −874 kJ/mol against −861 kJ/mol indicated by Robie et al., 1995) but in line with the assessment of Mao et al. (2001).

#### 5.1.2.2 Coesite

Stishovite transforms reversibly, with the decrease of P, into the less dense phase Coesite. For this polymorph, Kirfe et al. (1979) proposed the space group $P2_1/c$. More recent investigations (Smyth et al., 1987; Geisinger et al., 1987) assigned to Coesite the monoclinic space group $C2/c$. The DFT-LDA bulk modulus of Coesite, $K_0$ is 94.3 GPa with a P derivative $K' = 4.8$ (Demuth et al.,1999). These values are approximately consistent with the experimental observations of Levien and Prewitt (1981) and the assessment of Mao et al. (2001). In this work, we adopt $K_0 = 100.3$

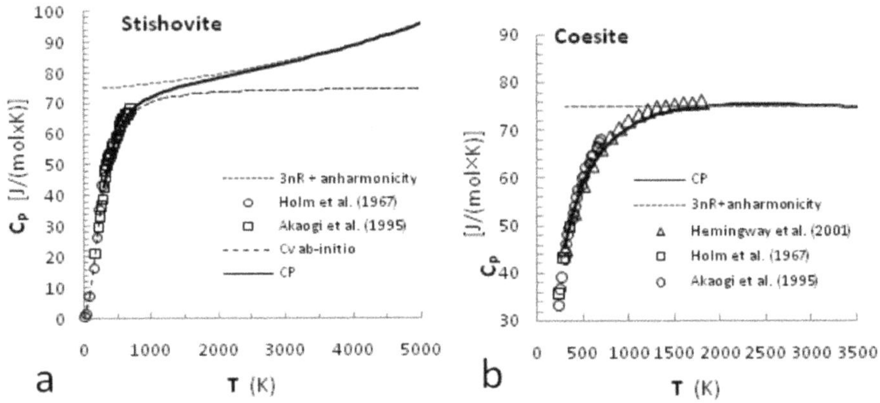


**Fig. 5.3** Isobaric and isochoric heat capacity of Stishovite (**a**) and Coesite (**b**), compared with low-T observations. The solid curve is the continuous function generated by Shomate's heat capacity expansion 4.1. The polynomial has been refitted to be valid at the high thermal regimes of interest in this study (Table 4.1). From Belmonte et al., 2017a, with modifications

GPa, $K' = 8.4$ (Fig. 5.1 and Table 4.2) coupled with a marked thermal dependence of the compressibility ($K_1 = -249.584$ bar/K; $K_2 = 0.0252$ bar/$K^2$). The adopted standard-state molar volume and the thermal expansion are based on the experiments of Bourova et al. (2006) (Fig. 5.4). For $C_P$, we refitted the experiments of Hemingway et al. (1998) in the form of a Shomate's equation consistent with the Dulong-Petit limit expectations at high-T (Fig. 5.3b). The anharmonicity contribution to the isochoric heat capacity (based on the above selected thermophysical parameters, and different from Stishovite) is almost negligible. The assessed values of S° and $H°_f$ are nearly similar to those of Mao et al. (2001).

### 5.1.2.3 Quartz

With the decrease of P, Coesite transforms reversibly into $\beta$ -Quartz for temperatures above ~1550 K and to $\alpha$-Quartz at lower T. The existence of a lambda transition coupled to a slight first-order contribution makes the assessment of the thermo-physical properties of the two Quartz polymorphs quite complex. Several Authors investigated the $\lambda$ transition of quartz (Helgeson et al., 1978; Hosieni et al., 1985; Salje et al.,1991; Berman and Brown, 1985, 1986; Berman, 1988), but the results, expressed as second-order contributions to the bulk thermodynamic properties of the phase, were somewhat contradictory. Adopting the procedure of Berman and coworkers, one assumes a common background (or "*lattice*" cf. Berman, 1988) heat capacity with superimposed $\lambda$-contributions. Thus, Belmonte et al. (2017a), based on the experiments of Ackermann and Sorrell (1974) and Bourova and Richet (1998), fixed the standard-state molar volume of $\beta$ -Quartz at 23.9 cc/mol and assumed the

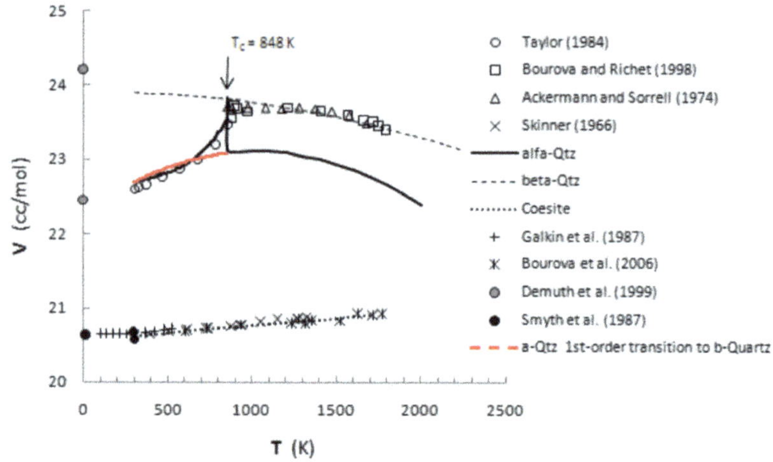


**Fig. 5.4** Model volumes of Quartz and Coesite, compared to experimental observations and first-principles computations. The red-dashed line depicts the hypothetical volume of the $\alpha$-polymorph in the absence of displacive (second-order) effects within the P = 1 bar transition zone. From Belmonte et al., 2017a, with modifications

existence of a small negative thermal expansion (i.e. $\alpha = -0.07 \times$ T; here retained; Table 4.2).

The $\alpha$-Quartz to $\beta$ -Quartz transition is in this way seen as a simple first-order event, with a constant Clapeyron slope ($k$ = dT/dP = 0.0237 K/bar). There is a $\lambda$ contribution to entropy at $T_C$ ($S_{\lambda,C}S$) of 2.158 J/[molxK)].

The volume at the critical temperature is $V_{\lambda,C} = 0.55$ cc/mol ($T_C \cong T_\lambda = 848$ K). Subtracting this value from the first-order volume of $\beta$ -Quartz, one obtains V° = 22.95 cc/mol, which is still somewhat too high with respect to the observation of Taylor (1984) (V° = 22.69 cc/mol; here retained).

The thermal evolution of the volume for $\alpha$-Quartz polymorph is sufficiently consistent with experiments (Fig. 5.4).

The DFT-LDA computations of Demuth et al. (1999) for the idealized $P6_222$ structure of $\beta$ -Quartz depict a marked stiffness of the substance at the athermal limit (i.e., $K_0 = 132.6$ GPa). The unique experimental observation of compressibility (i.e., 56.4 GPa at T = 873 K; Landolt-Börnstein, 1979) testifies to a marked T dependency, which, according to Demuth et al. (1999), is due to the activation of low-frequency Rigid Unit Modes (RUM).

Demuth et al. (1999) declare for the $P3_221$ $\alpha$—polymorph a DFT/LDA bulk modulus of 35.4 GPa at the athermal limit, to be compared with $K_0 = 45.0$ GPa of Hamann (1996). Mao et al. (2001), assign $K_0 = 40.07$ GPa and $K' = 6$. We adopt in this study a value nearer to Hamann's observation ($K_0 = 47.98$ GPa), while keeping $K'$ at 6.

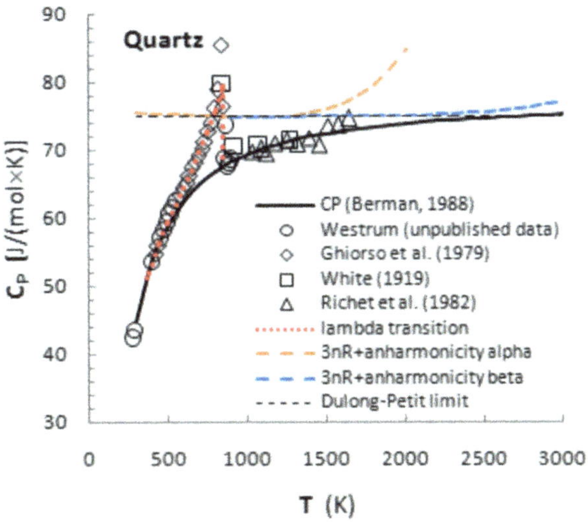


**Fig. 5.5** Isobaric and isochoric heat capacity of Quartz, compared with low-T observations. The effect of the second-order transition according to the Berman-Brown procedure (Berman and Brown, 1985,1986) is also shown

The adopted $K_1$ and $K_2$ are based on the experimental observations of the Clapeyron slope between the $\alpha$ polymorph and coesite made by Bose and Ganguly (2005) (Tables 4.2 and 4.3).

Already at relatively low-T, the $C_P$ of the $\beta$ phase is near the (Dulong-Petit) isochoric limit. This is indicative of low anharmonicity contributions for this polymorph, while it is not so for the $\alpha$ phase for which anharmonicity is quite relevant immediately after the phase transition (Fig. 5.5).

These complexities suggest extreme care in the extrapolation of the $\alpha$ phase beyond its extrinsic stability limit (i.e., $P \leq 3.2$ GPa; $T \leq 1500$ K; see later on).

#### 5.1.2.4 Cristobalite

Belmonte et al. (2017a) adopted for the $\alpha$-polymorph (space group $P4_12_12$) the V°, $K_0$, $K'$ values of Downs and Palmer (1994) (i.e., V° = 25.76 cc/mol; $K_0 = 11.5$ GPa; $K' = 9$). The experimental bulk modulus is well in line with the DFT/LDA predictions of Teter et al. (1995) and Demuth et al. (1999) (i.e., 11.9 and 12.8 GPa, respectively), though the baric derivative is much higher than the computed ones (3.0 and 6.3, respectively). For the $\beta$ -polymorph, Belmonte et al. (2017a) adopted bulk modulus and its baric derivative according to Demuth et al. (1999) for the $P2_13$ space group (i.e., $K_0 = 14.0$ GPa; $K' = 7.1$). The adopted standard-state volume (V° = 27.242 cc/mol) and thermal expansion of $\beta$ -cristobalite are consistent with the observations of Swainson and Dove (1995) and Bourova and Richet (1998) (Fig. 5.6 and Table 4.2).

Also, for cristobalite, the $\alpha/\beta$ transition is quite intriguing. To reproduce the experimental Clapeyron slope of transition, one must assume that the $\alpha$-polymorph has a complex thermal expansion (negative from room conditions to the $T_\lambda$ and positive

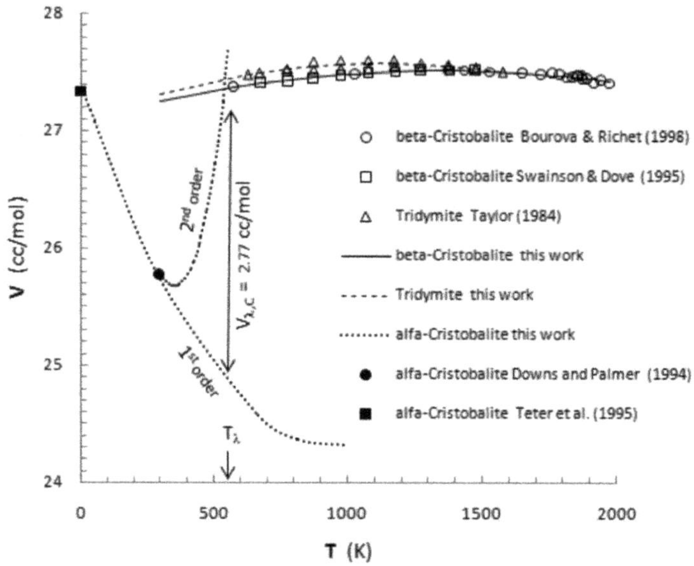


**Fig. 5.6** Model volumes of Cristobalite and Tridymite, compared to experimental observations. The effect of $\lambda$ displacive contributions to the bulk volume of the $\alpha$ polymorph is sketched. From Belmonte et al., 2017a, with modifications

at higher T; Belmonte et al. (2017a)). This complex behavior is consistent with both the computations of Teter et al. (1995) at the athermal limit and with the displacive contributions to volume at $T_\lambda$ ($\Delta V = 2.77$ cc/mol based on the Clapeyron slope; cf. Figure 5.6).

The $\lambda$ transition induces a substantial increase in heat capacity up to a critical limit that attains ~23 J/(mol × K) at $T_C = 550$ K (Mosesman and Pitzer, 1941). At this temperature, the delta in the bulk entropy of the two polymorphs is 2.276 J/(mol × K) and the $\beta$ polymorph attains at standard state an entropy of 46.029 J/(mol × K), consistent with Berman's (1988) assessment (Table 4.3 and Fig. 5.6). The "*lattice*" heat capacity function of Berman (1988) remains below the Dulong-Petit asymptotic limit in the T-range of extrinsic stability of the substance (Fig. 5.7).

#### 5.1.2.5 Tridymite

"*Tridymites*" occur in a variety of structural classes; hence, numerous phase transitions have been proposed for these low-density polymorphs of silica (Mosesman and Pitzer, 1941; Austin, 1954; Cohen and Klement, 1980; Raz, 1983). The ideal $C222_1$ orthorombic structure of $\beta$ -Tridymite is closely related to the structure of $\beta$ -cristobalite (Dmitriev et al., 1997). This high-T polymorph has been investigated by DFT/LDA procedures by Demuth et al. (1999), returning $K_0 = 30.8$ GPa and $K' = 4.1$.

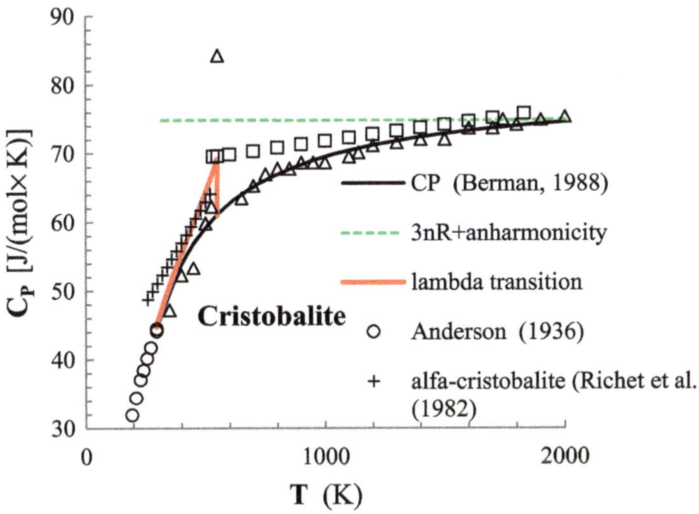


**Fig. 5.7** Heat capacity of Cristobalite, compared with low-T observations. The effect of the second-order transition from $\alpha$ ($D_{6h}$) to $\beta$ the polymorph is also shown

Based on the experiments of Taylor (1984), thermal expansion may be reproduced simply with two parameters (i.e., $\alpha_0 = -0.27655\text{E-}7$, $\alpha_1 = 313.204\text{E-}7$) and a standard-state molar volume $V^\circ = 27.306$ cc/mol (cf. Figure 5.6 and Table 4.2). Following Berman (1988), it may be assumed that only two-phase transitions occur in Tridymites, namely those for which "*acceptable consensus*" has been obtained: a low-to-mean transition ($T_\lambda = 388$ K) and a mean-to-high transition ($T_\lambda = 440$ K; Cohen, 1980; Raz, 1983). It may also be assumed for simplicity that the "*low*" and "*mean*" polymorphs are identical to the "*high*" polymorph (i.e., the ideal $C222_1$ orthorombic structure) in their thermophysical parameters.

Figure 5.8 shows the common *lattice* heat capacity function of Tridymite (Berman, 1988; Table 4.1), superimposed with the two-phase transitions. The second-order contributions of the two transitions (Table 4.3) at $T_C$ are the following: for the low/mean transition $\Delta H_\lambda = 473.4$ J/mol; $\Delta S_\lambda = 1.287$ J/(mol × K); $\Delta V_\lambda = 0.0$ J/bar and for the mean/high transition $\Delta H_\lambda = 91.5$ J/mol; $\Delta S_\lambda = 0.208$ J/(mol × K); $\Delta V_\lambda = 0.0136$ J/bar.

### 5.1.3 *Equilibria at Subsolidus*

As shown by Belmonte et al. (2017a) (Fig. 5.9), the Clapeyron slopes between neighboring Cristobalite and Tridymite polymorphs are reproduced faithfully within the limits of reversal experiments. It should be stressed for this purpose that, in the case of phase transition with a wide $T_{ref}$-$T_C$ gap and affected by a marked Clapeyron

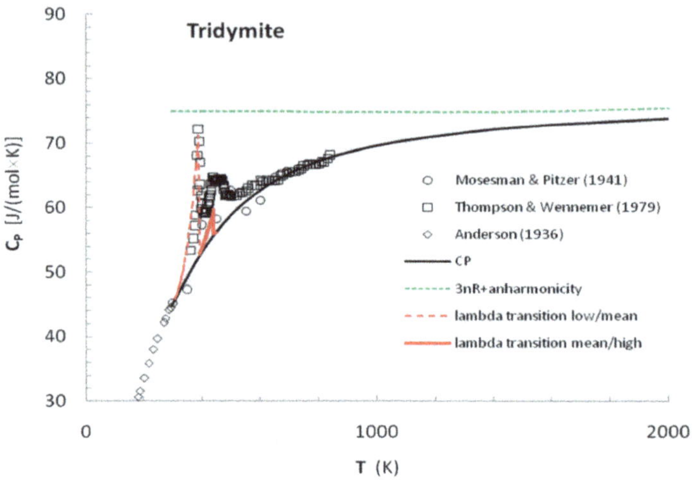


**Fig. 5.8** Heat capacity of Tridymite, compared with low-T observations. The effect of the second-order low/mean and mean/high transitions is also shown

slope (such as for the case of cristobalite), the way of treating the phase transition thermodynamics affects the depicted extrinsic stability limit between the two polymorphs.

In the adopted model at room pressure the $\lambda$-transition begins at $T_{ref} = 298.15$ and ends at $T_C = T_{C,\lambda} = 550$ K. Ignoring the $\lambda$ contributions to the stability of the $\alpha$ cristobalite polymorph would predict a first-order phase transition already at T = 522 K, while, allowing the $T'_{C,\lambda}$ to migrate with pressure along the specified dT/dP slope (i.e. 0.1217 K/bar; cf. Table 4.3) the transition at room pressure takes place correctly at $T_{C,\lambda} = 550$ K but the Clapeyron slope is somewhat affected (i.e. dT/dP

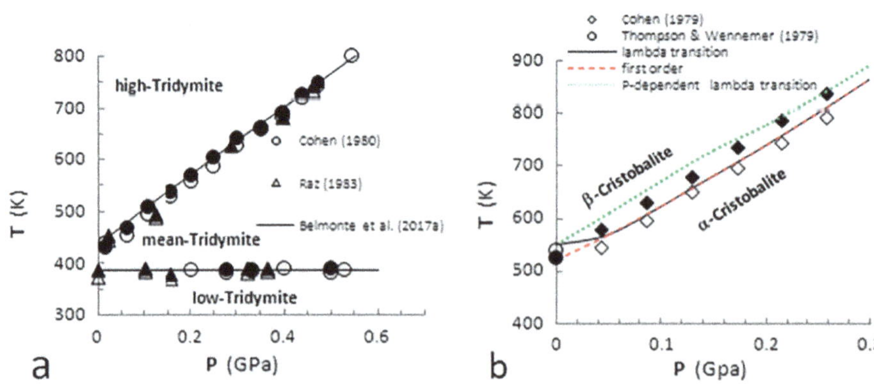


**Fig. 5.9** Phase transitions in Tridymites (a) and Cristobalite (b). The results of the present investigation are compared with experimental observations. Full and empty symbols denote reversals. From Belmonte et al. (2017a) with modifications

= 0.1131 k/bar). To avoid this misbehavior (virtually negligible for the other phases considered in this study), one should perform a complex $V_P dP$ integration along the entire transition boundary. In practice, this is not necessary: the transition may be treated as first-order (embodying the $\lambda$-contributions at the standard state into the H°, S°, V° values). The inspection of the second-order effect is, however, important in depicting the correct thermal expansion of the low-T polymorph, as previously discussed.

The assessment of the low-density polymorphs of silica and of the pure liquid allows us to depict the phase relations at subsolidus in the low-P range as well as their melting curve. The results of this exercise are shown in Fig. 5.10. The agreement with experimental observations is reasonable. The $\beta$ -Crystobalite/$\beta$ -Quartz/liquid invariant point occurs, according to our assessment at P = 0.48 GPa and T = 1990 K, while the high-Tridymite/$\beta$ -cristobalite/$\beta$ -Quartz invariant point is localized at P = 0.152 GPa and T = 1439 K (P = 0.151 GPa and T = 1440 K according to Berman, 1988).

In Fig. 5.11, the phase diagram of $SiO_2$ in the 1 $\leq$ P (GPa $\leq$ 6 range is drawn. The experimental observations concerning the univariant boundary between the two Quartz polymorphs and between $\beta$ -Quartz and coesite are sufficiently well reproduced with respect to the observations of Cohen and Klement (1967) and Mirwald and Massonne (1980,1982). The $\alpha$-Quartz/coesite boundary is apparently a simple system for experimental study. Bose and Ganguly (2005) carried out a detailed investigation of friction correction terms in the piston-cylinder apparatus, and their result/

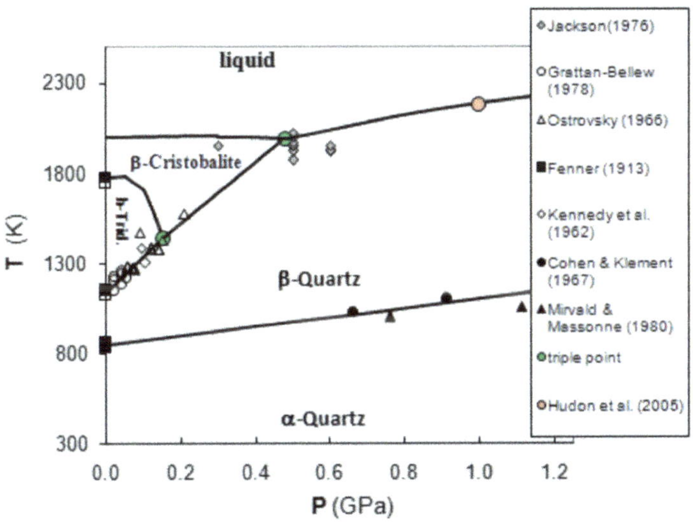


**Fig. 5.10** $SiO_2$ phase diagram below 1.25 GPa. Liquidus and subsolidus extrinsic stability limit between Cristobalite, high-Tridymite, and the $\alpha$, $\beta$ forms of Quartz obtained in this study (solid lines) are compared with experimental observations

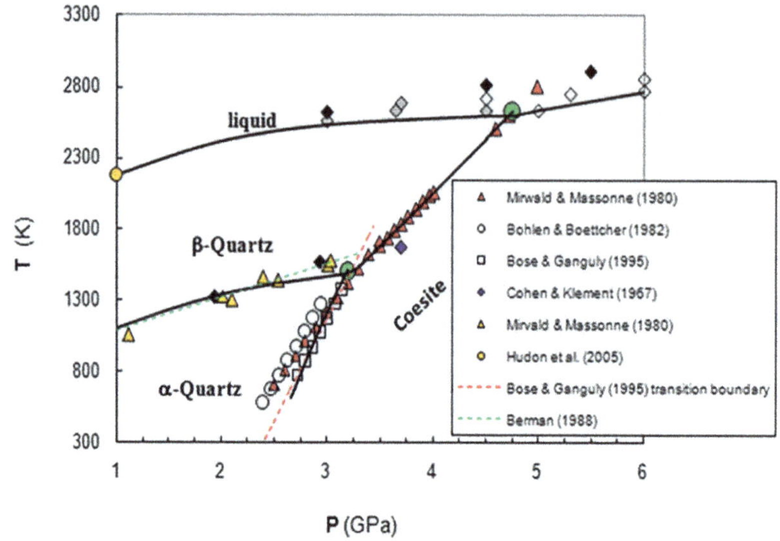


**Fig. 5.11** $SiO_2$ phase diagram at low pressure ($P \leq 6$ GPa). Liquidus and subsolidus extrinsic stability limit between Coesite and the $\alpha$, $\beta$ forms of Quartz obtained in this study (solid lines) are compared with experimental observations

analysis has now been widely accepted as the benchmark. We selected their transition boundary to refine the thermal dependence of the $\alpha$ polymorph.

The $\beta$ -Quartz/Coesite/liquid invariant point is predicted to occur at P = 4.75 GPa and T ≅ 2630 K. The $\alpha$-Quartz/$\beta$ -Quartz/coesite invariant point is observed at P = 3.2 GPa and T = 1514 K.

The univariant coesite/Stishovite/Liquid boundary reproduces satisfactorily the observations of Akaogi et al. (1995) and Zhang et al. (1993). The invariant point is predicted to occur at P = 13.8 GPa and T = 3173 K (Fig. 5.12). As anticipated, the melting curves of Stishovite and coesite are consistent with the observations of Kanzaki (1990) and Shen and Lazor (1995).

## 5.2 $Al_2O_3$

### 5.2.1 $Al_2O_3$ Liquid

At the melting point, $\alpha$-alumina undergoes a dramatic structural reorganization which does not involve simple rupture of bonds subjected to thermal stress, but deep rearrangements in short-range ordering with a substantial change in the coordination of the central cation. The density deficit between crystalline and amorphous states

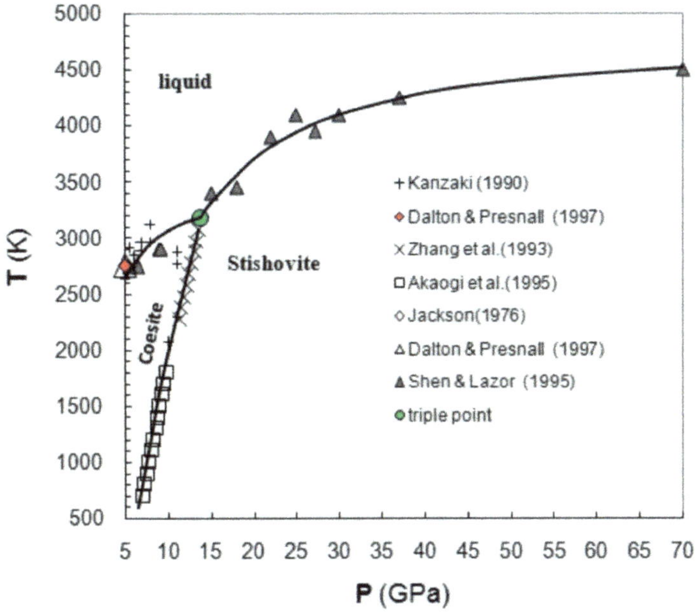


**Fig. 5.12** $SiO_2$ phase diagram at high pressure (5–70 GPa). Liquidus and subsolidus extrinsic stability limit between Stishovite and Coesite obtained in this study (solid lines) are compared with experimental observations. From Belmonte et al. (2017a), with modifications

attains in this way 27% (Landron et al., 2001). Although the heat capacity of $\alpha$-alumina has been the object of intensive investigation (Castanet, 1984, and references therein), only a few experimental studies are available on the liquid (Kantor et al., 1962; Shpil'rain et al., 1962). The properties of liquid alumina have been investigated by Belmonte et al. (2013) through first principles, with the same procedures adopted for liquid silica. Based on the computed energies and vibrational features, an aggregate of the *D3d6* positively-charged cluster $[Al_{12}O_{11}]^{14+}$ contoured by $[AlO_4]^{5-}$ units in a dielectric continuum with dielectric constant $\varepsilon = 4.575$ an approximate 1:3 proportion generates an isobaric heat capacity which is, at high-T, roughly consistent with the estimates of Castanet (1984) and, at intermediate T, with the assessment of Stebbins et al. (1984) (within error bars). Also, in this case, the apparent heat capacity gap in the homogeneity range of the substance depicted at $T_f$ by controversial literature estimates is hard to justify unless we admit sudden and substantial rearrangements of the amorphous undercooled liquid with respect to the stable state (Fig. 5.13).

The thermal expansion of the liquid was obtained by Belmonte et al. (2013) through inverse modeling of the melting curve of the substance (see later). The average value of the assessed thermal expansion coefficient in the T range from 2000 to 3100 K (i.e., $\alpha_V = 4.9 \times 10^{-5}\ K^{-1}$) is slightly higher than those measured by electrostatic or aerodynamic levitation techniques, ranging from $3.4 \times 10^{-5}$ to

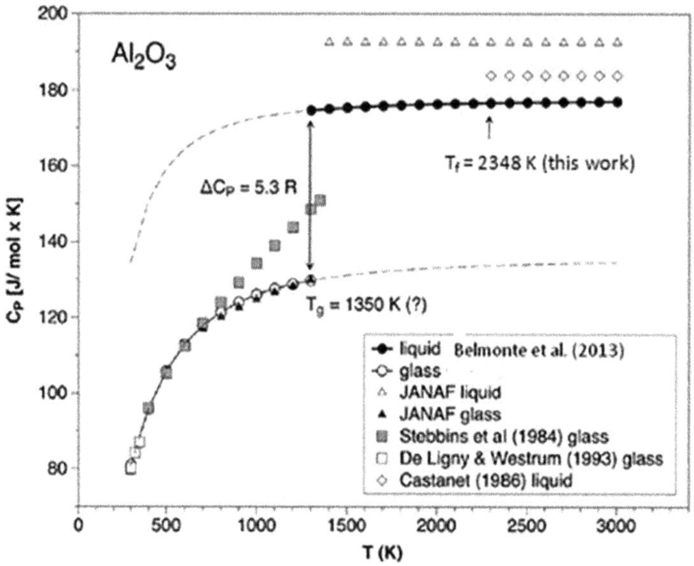


**Fig. 5.13** Calculated vs. observed isobaric heat capacity of amorphous $Al_2O_3$ at high-T. The "liquid" is seen as an aggregate of $[Al_{12}O_{11}]^{14+}$ islands coulombically contoured by $[AlO_4]^{5-}$ units in an approximate 1:3 proportion (actually 0.263:0.737). Coherent motions are missing, and anharmonicity is almost negligible. The effect of the translational and rotational partition function corresponding to 4.55R is added to the vibrational part (the structural subunits are free to rotate and translate in the dielectric medium; cf. Ottonello et al., 2010a). The glass heat capacity is obtained by summing the 219 optical modes of the $[Al_{30}O_{45}]^0$ cluster (resolved at the simple HF/STO-3G level of theory), the acoustic phonons, and the coherent excess mode, after scaling to the $Al_2O_3$ stoichiometry. The glass transition is not known with precision, but the heat capacity gap between the superheated glass and liquid is almost constant (from 5.3R at T = 1350 K to 5.2R at T = 2350 K). Modified from Belmonte et al. (2013)

$4.3 \times 10^{-5}$ $K^{-1}$ according to different experimental works (Glorieux et al., 1999; Paradis et al., 2004; Langstaff et al., 2013). Nevertheless, this discrepancy seems to be reasonable if compared with $\alpha_V$ values measured for liquid alumina by conventional container-based techniques (e.g., Kirshenbaum and Cahill, 1960), which are overestimated by an order of magnitude due to sample contamination effects.

### 5.2.2 α-$Al_2O_3$ (Corundum)

Pure $\alpha$-alumina (corundum) was promoted in 1948 as a heat capacity standard because it does not experience any solid-state transformation from the athermal limit to the melting point, which, at room pressure, takes place at ~2345 K (2348 K in our assessment). $\alpha$-alumina [*D3d6* (*R-3c*)] unit was investigated by Belmonte et al. (2013), within the LCAO approximation. Calculations were carried out with the Density

Functional Theory (DFT) and the hybrid B3LYP functional. The adopted basis set corresponds to the 85-11G(*d*) and 8-411G(*d*) contractions for Al and O adopted by Montanari et al. (2006) in their *ab-initio* investigation of the vibrational partition function of corundum. The adopted computational parameters were, however, more stringent. The level of accuracy in evaluating the Coulomb and Hartree–Fock exchange series is, in fact, controlled by five parameters (Dovesi et al., 2009) that were set to 9, 9, 9, 9, 18. The threshold of the SCF energy was set to $10^{-8}$ hartree for the geometry optimization and $10^{-10}$ hartree for the vibrational frequency calculations. The reciprocal space was sampled according to a regular sublattice with a shrinking factor equal to 6 and corresponding to 32 independent *k*-points in the irreducible part of the Brillouin zone. The result of this computational exercise is virtually indistinguishable from the calorimetric assessment within the experimental T-range of observation (Ditmars et al., 1982) (Fig. 5.14).

We adopt here for alumina the enthalpy of formation from the elements at standard state of Berman (1988) (i.e., $H^0_{f,298.15} = -1675.7$ kJ/mol; Table 4.1). However, we adopt for $S^0_{298.15}$ the value obtained *ab-initio* by Belmonte et al. (2013), which is substantially higher than Berman's value (i.e., 51.4 against 50.82 J/(mol × K)]. Adopting $T_f = 2325$ K, the enthalpy of fusion, as determined from the difference between the heat capacity delta of $\alpha$-alumina from the melting point to the athermal

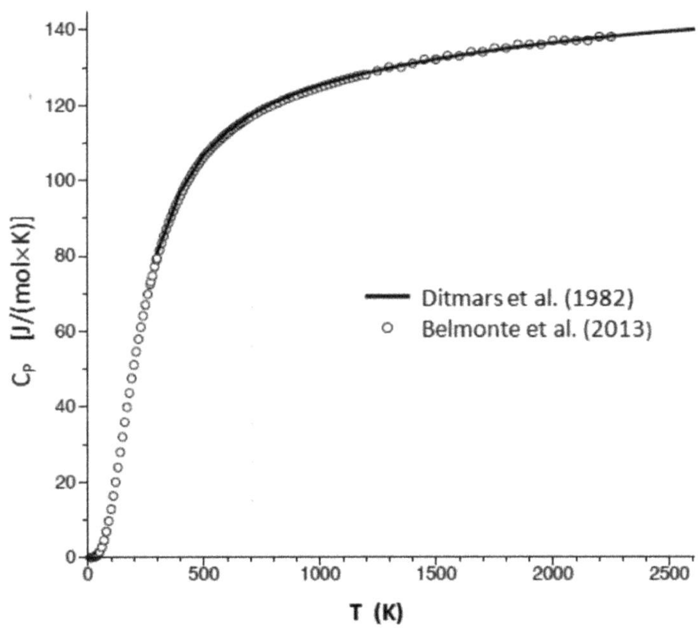


**Fig. 5.14** Ab-initio bulk isobaric heat capacity of $\alpha$-$Al_2O_3$ (Belmonte et al., 2013; solid line) compared with smoothed values arising from calorimetric measurements on the Standard Reference Material 720 (Ditmars et al., 1982). The bias in the T-range from room conditions to the melting temperature does not exceed one J/(mol × K) in most cases. Modified from Belmonte et al. (2013)

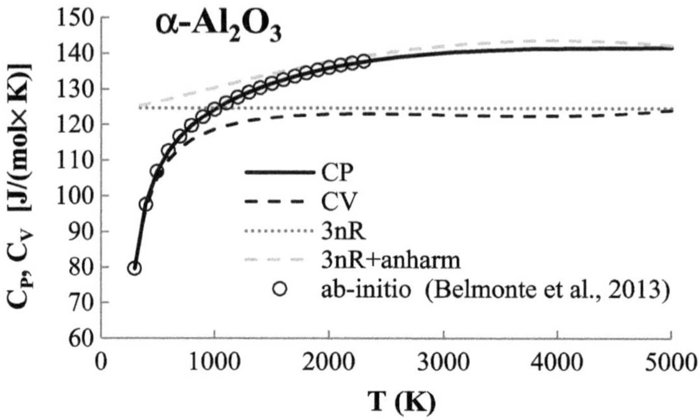


**Fig. 5.15** Heat capacity of $\alpha$-$A_2O_3$ (corundum) expanded to high-T by refitting of Shomate's coefficients previously obtained by Belmonte et al. (2013) on the basis of their *ab-initio* values. This new assessment allows a further optimization of the thermoelastic properties of the substance, constrained to obey the Dulong-Petit limit

limit ($H\alpha_{,2325}$—$H\alpha_{,0}$) and the corresponding value for the liquid ($H_{L,2325}$—$H_{L,0}$), is 107 ± 5 kJ/mol (Castanet, 1984; Shpil'rain et al., 1962). The *ab-initio* heat capacity of $\alpha$-$Al_2O_3$ calculated by Belmonte et al. (2013) is in excellent agreement with the calorimetric assessment of Ditmars and Douglas (1971). It has been extended to higher T in this study by refitting Shomate's coefficients (Table 4.1), and it is now valid up to 5000 K (Fig. 5.15). It was thus possible to re-assess the thermoelastic properties of the substance in order to obey the Dulong-Petit limit at high-T. To achieve this goal is was sufficient to decrease $K_1$ from −390.4 bar/K (Belmonte et al., 2013) to −350 bar/Km, keeping all the other parameters unvaried.

The melting behavior of $\alpha$-$Al_2O_3$ is poorly known. The single LHDAC determination (Shen and Lazor, 1995) is in agreement with the MD simulation of (Ahuja et al., 1998), indicating a melting temperature $T_f = 3500 \div 3750$ K at P = 25 GPa. Shen and Lazor (1995) adopted a melting temperature $T_{fusion} = 3525 \pm 150$ K at P = 25 GPa, in agreement with MD (Ahuja et al. 1998). The melting curve has been reproduced here through an assessment of the thermophysical properties of $\alpha$-$Al_2O_3$ (Fig. 5.16; see also Table 4.2), stemming from the optimized data of $\alpha$-$Al_2O_3$ of Berman et al. (1985) and the thermophysical assessment of the liquid after (Belmonte et al. 2013). The computational results are not far from what was anticipated by Ottonello et al. (2013) (see their Fig. 2.2). The agreement with the experimental observations and MD simulations is acceptable.

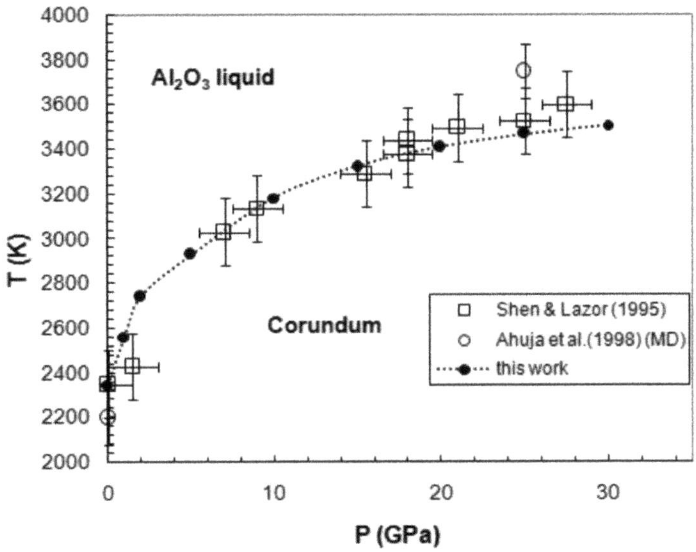


**Fig. 5.16** Melting curve of corundum ($\alpha$-$Al_2O_3$). The melting temperature is slightly underestimated at high P but still within the error bars of the LHDAC measurements of Shen and Lazor (1995). At room conditions, we reproduce satisfactorily the observation of Shen and Lazor (1995), i.e., 2350 K, somewhat higher than the previous estimates (2327 K) (see Fig. 5.13)

### *5.2.3 MgO*

### *5.2.4 MgO Liquid*

The assessment of the melting behavior of the pure component MgO is flawed by a poor and conflicting knowledge of its heat capacity, molar volume, and thermal expansion in the liquid aggregation state. Also, the melting behavior of the unique crystalline polymorph that nucleates in the wide P range of interest in this study is poorly known. The heat capacity of the liquid above the melting temperature is assumed in the NIST-JANAF tables (Chase, 1998) to be 16 cal/(mol × K) simply in analogy to what is observed for liquid BeO [$C_P \geq 19.4$ cal/(mol × K)] and to attain in the super-cooled liquid values actually observed for periclase. The same tables adopt the melting temperature of MgO at room pressure, a compromise value of 3105 K (McNally et al., 1961), and place the glass transition arbitrarily at roughly 2/3 of this value ($T_g = 2100$ K). The enthalpy of melting is quite uncertain, and the NIST-JANAF guess value is $\Delta H_f = 77.8$ kJ/mol. The corresponding $\Delta S_f$ is 25.1 J/(mol × K), which is not far from the value obtained by Alfé (2005) through first principles [$\Delta S_f \cong 16.1$ J/(mol × K)]. As observed by Karki et al. (2006) through First Principles Molecular Dynamics calculations (FPMD) carried out in the frame of Density Functional Theory (DFT) with pseudo-potentials, liquid MgO undergoes progressive structural changes with varying T and P. At low pressure, the Radial Distribution function depicts a Mg-O first neighbor coordination number of around 4.5. This value rises progressively to 7 with the increase in pressure. The observations of Karki et al. (2006) prompted us to attempt a simple DFT-PCM vibrational calculation of tetrahedral clusters in a medium with permittivity $\varepsilon = 13.3$ and an appropriate solvation radius (Ottonello and Richet, 2014). It was then assumed that the isochoric heat capacity of the liquid at P = 1 bar and 3000 K attains the value obtained by de Koker et al. (2013) through FPMD computations and decreases according to the contributions of the nine normal modes of the tetrahedral cluster as calculated ($\nu_1 = 238\ \text{cm}^{-1}$; $\nu_{2,3} = 243\ \text{cm}^{-1}$; $\nu_{4,5} = 267\ \text{cm}^{-1}$; $\nu_6 = 270\ \text{cm}^{-1}$; $\nu_{7,8,9} = 449\ \text{cm}^{-1}$) and by the effect of anharmonicity. As we may see in Fig. 5.17, the lowering of $C_P$ with the decrease of T mimics what is observed in periclase (Belmonte, 2013; Barin, 1995), but it implies that rotational and translational components corresponding to roughly $1.6 \times R$ are present in the liquid and cease out at the glass transition.

Transforming the isochoric heat capacity into its isobaric counterpart requires knowledge of the anharmonic contributions. The thermoelastic parameters of the pure liquid may be optimized on the basis of the melting curve of the substance along the guidelines discussed in Ottonello et al. (2010a, 2013). The substance is initially assumed to behave in a strictly harmonic way and to reproduce the univariant loci experimentally observed at various P,T conditions. Initial guess values of the standard-state molar volume (fictive liquid at T = 298.15 K and P = 1 bar) bulk

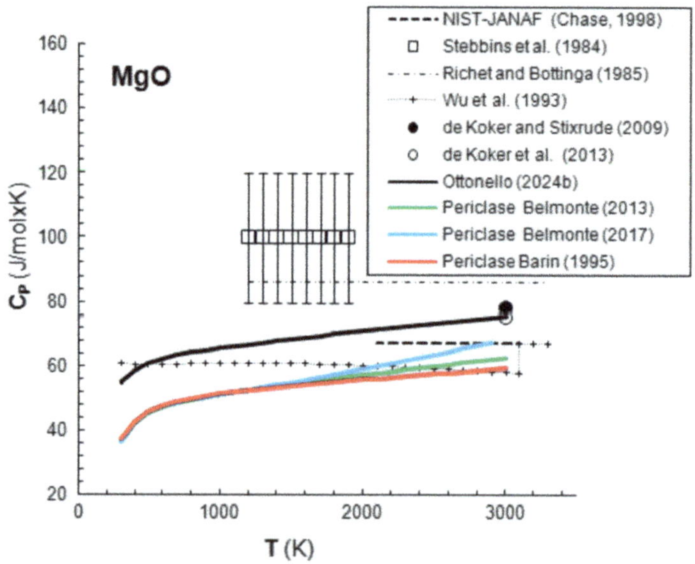


**Fig. 5.17** Isobaric heat capacity of MgO liquid, compared with various assessments (Chase, 1998; Richet and Bottinga, 1985; Wu et al., 1993; Stebbins et al., 1984) and results of FPMD computations (de Koker et al., 2013; de Koker and Stixrude, 2009). The isobaric heat capacity of periclase (Belmonte, 2013,2017; Barin, 1995) is also shown for comparative purposes

modulus at the athermal limit ($K_0$), its baric and thermal dependances ($K'$, $K_1$, $K_2$) and of the thermal expansion (represented as a polynomial on T of type $\alpha_V = \alpha_0 T + \alpha_1 + \alpha_2 T^{-1} + \alpha_3 T^{-2} + \alpha_4 T^{-3}$) are necessary. The anharmonic contributions generated by the inverse procedure are then retrieved in terms of a, c, f, and g coefficients of the Shomate's Eq. 4.1 and added to the $C_V$ equation. Calculations are iteratively refined to attain univariant loci consistent with the experimental observations. The results of this exercise are summarized in Tables 4.1 and 4.2. The standard-state enthalpy of formation from the elements and entropy of MgO in the liquid aggregation state were obtained by adopting the NIST-JANAF estimates for the latent heat of crystallization.

In Fig. 5.18, the melting curve of MgO depicted by *ab-initio* calculations is compared with the few experimental data at disposal and of MD and FPMD investigations. Here again, we must note that the literature is quite controversial and hundreds of K separate the high-P melting temperature of the substance calculated through FPMD and MD methods from experiments. Moreover, the role of anharmonicity is substantial, due to a marked thermal expansion, and renders the results somewhat path dependent [14.6 J/(mol × K) of anharmonicity at 3000 K in Ottonello (2024a, 2024b); 18.3 J/(mol × K) according to de Koker and Stixrude (2009) and 15.0 according to de Koker et al. (2013) at the same T; a maximum contribution attaining roughly 30% at high-T]. Having adopted the isochoric heat capacity of de Koker et al. (2013) as a pivotal point to determine the provisional contributions of $Q_{tra}$ and $Q_{rot}$ and having adopted the bulk modulus $K_0$ and $K'$ of Karki et al. (2006), it is possible to refine the thermal expansion in such a way as to attain at the mantle/core boundary (135.6 GPa) melting temperatures consistent with the results of their FPMD investigation (Ottonello, 2024a, 2024b). The obtained melting curve fits nicely with the recent experiments of Du and Lee (2014) and the computations of Belonoshko et al. (2010) in the low to intermediate-pressure regimes, while attaining melting temperatures more consistent with the calculations of Boates and Bonev (2013) and de Koker et al. (2013) at higher pressure regimes.

The resulting $P = 1$ bar molar volumes of liquid MgO are intermediate between the assessment of Bottinga et al. (1983) and the FPMD estimates of de Koker et al. (2013). There is, particularly at intermediate T, a close agreement with the experimental observations of Ikada et al. (1967) and Tomlinson et al. (1958) (Fig. 5.19).

### 5.2.5 *Periclase*

MgO crystallizes only in the cubic form *Fm3m* (*Periclase, fcc* or B1-type structure), and no phase transitions are known to occur in the wide P–T range consistent with planetary regimes.

The first *ab-initio* computation of the elastic constants and the high symmetry point (Γ,, X, and L) phonon frequencies of MgO has been carried out by McCarthy and Harrison (1994) at the HF theory level. In that study, correlation-only corrections to the ground-state periodic HF energies were added *a-posteriori*. The most

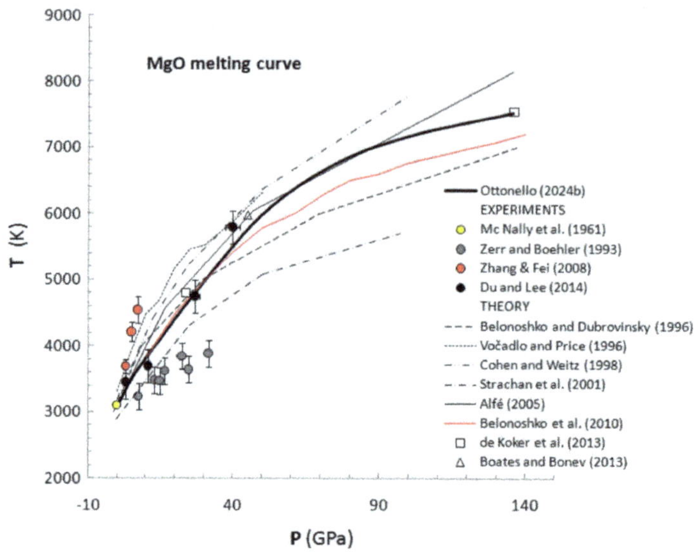


**Fig. 5.18** Melting curve of MgO in a wide pressure range, compared with experiments and results of semi-empirical and first-principles MD calculations. From Ottonello (2024b) with modifications

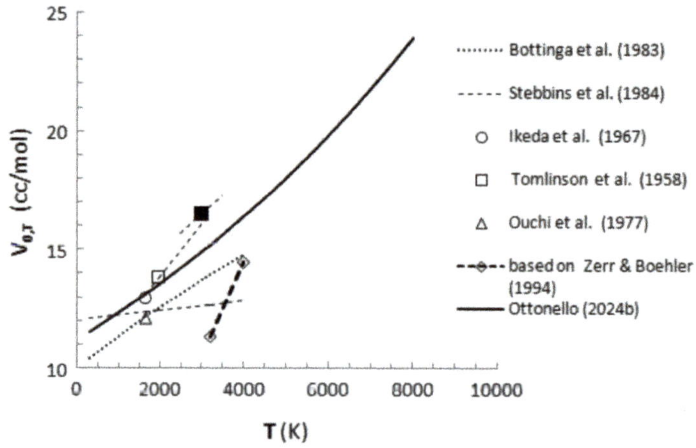


**Fig. 5.19** Isobaric (P = $10^5$ Pa) molar volume of MgO. Values derived from the experimental observations of Tomlinson et al. (1958), Ikeda et al. (1967), and Ouchi et al. (1977) in binary and ternary systems at T = 1400 °C compared with the FPMD estimates of De Koker et al. (2013) and the HPA assessment. From Ottonello (2024b), with modifications

comprehensive *ab-initio* investigation hitherto appeared of thermophysical and thermodynamic properties of Periclase is due to Belmonte (2017). Computations were carried out at the DFT/B3LYP theory level under GGA and LCAO approximation. Magnesium and oxÿgen were described by (8 s)-(511sp)-(1d) and (8 s)-(411sp)-(1d) contractions for Mg and O respectively, with $\alpha_{sp} = 0.2245$ and $\alpha_d = 0.50$ exponents for the most diffuse functions of Mg (in $Bohr^{-2}$) and $\alpha_{sp} = 0.181$ and $\alpha_d = 0.60$ for O. MgO displays, at high P – high-T conditions, a strong single-crystal elastic anisotropy (Karki et al., 1997; Tsuchiya and Nakamura, 2001). Moreover, some aspects related to its thermodynamic and thermophysical properties under those conditions are still poorly defined (Wentzcovitch et al., 2010). This evidence pushed Belmonte (2017) to extend the first-principle investigation to very high P and T (i.e., 0–160 GPa; 0–3000 K). Besides the usual vibrational frequencies at Γ point (i.e., $\mathbf{k} = 0$), Belmonte (2017) performed phonon dispersion calculations at many other special points with a direct-space approach (Parlinski et al., 1997; Alfé, 2009) on a $5 \times 5 \times 5$ supercell with 250 atoms and 750 normal modes sampled at 125 k-points. This ensured convergence on harmonic thermodynamic functions (e.g., isochoric heat capacity and vibrational entropy) in the whole P–T range. Based on the investigation of Belmonte (2017), the $\alpha_T K_T$ product of MgO attains at high-T a constant value of roughly 65 bar/K while both experiments (Barron et al., 1959; Dubrovinsky and Saxena, 1997) and assessments (Anderson and Zou, 1990; Singh et al., 2004) seem to indicate that $\alpha_T K_T$ should decrease above the Debye temperature ($\theta_D$).

It should be noted that these purposes that the attainment of an asymptotic limit at high-T in the $\alpha_T K_T$ product is implicit in the functional form of the equation, based on first principles:

$$\alpha_T K_T(V, T) = \frac{R}{zV} \sum_{\mathbf{k},i=1}^{3n} \gamma_i(\mathbf{k}, V) e^{x_i} \left( \frac{x_i(\mathbf{k}, V)}{e^{x_i(\mathbf{k},V)} - 1} \right)^2 \tag{5.1}$$

**k** in (5.1) are the investigated special points, $Z$ are the formula units in the primitive unit cell. The summation in (5.1) is extended not only to all optical modes in the first Brillouin zone but to the acoustic branches as well, and the notation emphasizes that the a-dimensional frequencies ($x$) are subjected to phonon dispersion (besides being V-dependent). Moreover, the so-called $\gamma$ is given by the ratio:

$$\gamma_i = -\frac{\partial \ln \nu_i}{\partial \ln V} \tag{5.2}$$

where $\nu$ are the linearized frequencies.

The thermal dependency of the bulk modulus is obtained by applying (Anderson, 1995):

$$\left( \frac{\partial K}{\partial T} \right)_P = -\overline{\alpha} \overline{K} \times \left[ K' - q^{ht} + 1 \right] \tag{5.3}$$

where $\overline{\alpha K}$ is the limiting αK product at $T \rightarrow \infty$

$$\overline{\gamma}_{ht} = \left(\frac{1}{3n-3}\right)\sum_{i=4}^{3n}\gamma_i \tag{5.4}$$

$$q^{ht} = \frac{\ln \overline{\gamma}_{ht} - \ln \gamma_0}{\ln V_{ht} - \ln V_0} \tag{5.5}$$

$\gamma_0$ in Eq. (5.5) is the mode gamma at the athermal limit, i.e., that obtained *ab-initio.* Values of γ and q may be obtained experimentally only by adiabatic decompression methods (Boehler and Ramakrishnan, 1980; Boehler, 1982). There is, however, a strict interdependence between $\gamma_0$ and q[1] (see also Anderson, 1974), and, setting $\gamma_0$ to zero, one has

$$q^{ht} \cong \frac{\ln \overline{\gamma}_{ht}}{\ln\left(V_{ht}/V_0\right)} \tag{5.6}$$

By knowing the $\alpha_T K_T$ product and the thermal effect on K, one may retrieve the thermal expansion at all *T* conditions and recast it eventually as a polynomial over reciprocal temperature with negligible error in reproducibility.

Belmonte (2017) obtained $(dK/dT)_{P=0}$ and $K''$, through full phonon dispersion calculation of P–V–T relations fitted with a BM3-EOS. His values are in agreement with the experimental results determined by rectangular parallelepiped resonance technique in the range T = 300–1800 K at P = 1 bar (Isaak et al., 1989). However, the thermal expansion obtained by Belmonte (2017) is overestimated at high-T and deviates consistently from observations (Dubrovinsky and Saxena, 1997) above 2000 K. As a net effect, the anharmonicity of periclase is also overestimated, resulting in a $C_P = f(T)$ functional inconsistent with the NIST-JANAF tabulation (Barin, 1995; Chase,1998) but nearer to FPMD predictions (de Koker and Stixrude, 2009; de Koker et al., 2013; Fig. 5.17).

### 5.2.6 CaO

### 5.2.7 CaO Liquid

As observed by Bajgain et al. (2015) through First Principles Molecular Dynamics calculations (FPMD) carried out in the frame of Density Functional Theory (DFT) with pseudo-potentials, liquid CaO undergoes progressive structural changes with

---

[1] "*we apply the frequently used power law relationship* $\gamma = \gamma_0(V/V_0)^q$ (3.2), *where* $\gamma_0$*is* γ *at zero pressure. Using* (3.2), *various sets of* $\gamma_0$*and q are chosen, and* γ *values are calculated at pressures where dT/dP measurements were carried out*" (Boehler and Ramakrishnan, 1980).

varying T and P. At low pressure, the Radial Distribution function depicts a Ca-O first neighbor coordination number around 5. This value rises progressively to 8 with the increase in pressure. According to the same Authors, the isochoric heat capacity of liquid CaO at 3000 K is $C_V = 44.90 \pm 1.83$ J/(mol × K). Based on the listed thermophysical parameters at the same temperature, the anharmonicity correction is $TV\alpha^2K = 21.77 \pm 2.75$ J/(mol × K), which, when added to the isochoric value, gives an isobaric heat capacity $C_P = 66.67 \pm 4.58$ J/(mol × K). This value is, within error, consistent with the NIST-JANAF estimates, which assign to liquid CaO a constant heat capacity of 62.76 J/(mol × K). The melting temperature at room pressure is quite controversial. As emphasized by Eriksson et al. (1994), the oldest assessments agreed in locating the melting point of CaO at $T_f = 2868 \pm 35$ K (cf. Elliott and Gleiser, 1960). This value is accepted in many thermodynamic assessments (e.g. Eriksson and Pelton, 1993; Wu et al., 1993; Eriksson et al., 1994; Jung et al., 2004, 2005). More recent investigations, however, substantially raised the 1 bar melting temperature (more precisely $T_f = 3200 \pm 50$ K according to Traverse and Foex, 1969; $T_f = 3172$ K according to Yamada et al., 1986). A recent experimental investigation by Manara et al. (2014) confirms this higher melting temperature of lime (i.e., $T_f = 3222 \pm 25$ K at 1 bar pressure). A recent Molecular Dynamics simulation carried out by Sun et al. (2010) adopts $T_f = 3200$ K and suggests a marked increase in the T of melting with pressure (see later). According to Kubaschewski et al. (1967), the heat of fusion is $\Delta H_f = 79.496$ kJ/mol. The corresponding entropy of fusion depends obviously upon the adopted $T_f$ [i.e., $\Delta S_f = 24.843$ J/(mol × K) at 3200 K (Barin, 1995) or 27.942 J/(mol × K) at 2845 K (Eriksson et al., 1994)]. Also, the heat capacity gap at $T_f$ is controversial: quite small but positive when one adopts the heat capacity function of Berman et al. (1985) for solid lime and the NIST-JANAF estimates for the liquid and negative [i.e., −0.672 J/(mol × K) at 3200 K] based on Barin (1995) tabulation. The heat capacity gap at $T_f$ is crucial in understanding the complex vibrational behavior of a substance at its melting point. In fact, as already outlined for other molten components, to a sudden increase of vibrational freedom (rotational and translational components) corresponds to the loss of coherent movement (i.e., acoustic motions) and also the anharmonic contributions may change substantially (Ottonello et al., 2010a; Belmonte et al., 2017a). The vibrational heat capacity of CaO arising from the three optical modes of the CaO asymmetric unit under orthorhombic symmetry was resolved ab-initio at the DFT theory level with a hybrid functional (B3LYP) and a 6-31G(*d,p*) basis set with the CRYSTAL package (Dovesi et al., 2014). When combined with the anharmonic contributions obtained by inverse methods from the analysis of the CaO melting curve, and with the 3R contributions arising from complete translational freedom of the atoms of the unit, the resulting isobaric heat capacity practically overlaps with the FPMD predictions of Baigain et al. (2015) (Fig. 5.20). The standard-state values of the liquid listed in Table 5.1 ($H^\circ_f$, $S^\circ$; pure component a $T = 298.15$ K, $P = 10^5$ Pa) were obtained by thermochemical cycle calculations assuming CaO to melt at 3222 K. Based on our calculations the entropy of fusion is ~7 cal/mol [$S_f = 29.207$ J/(mol × K)], slightly larger than that of MgO (and in agreement with the fact that the former compound is more refractory). The bulk modulus of CaO liquid and its thermal derivative, here

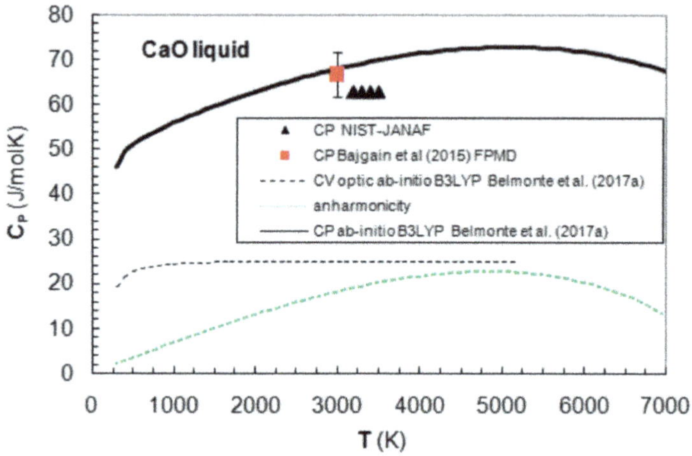


**Fig. 5.20** Ab-initio isobaric heat capacity of CaO in the liquid aggregation state. The isochoric vibrational contribution is generated by the optical phonons $\nu_{1,2} = 275.953$ cm$^{-1}$; $\nu_3 = 552.391$ cm$^{-1}$. The anharmonic contribution is given by the $VT\alpha^2K$ product, with thermal expansion $\alpha$ and bulk modulus K obtained by inverse methods from the thermodynamic analysis of the melting curve of the substance. To these contributions, the effect of complete translational freedom of the asymmetric unit is added to obtain the bulk isobaric heat capacity. Nonlinear minimization of the discretized values in the 298.15—7000 K T-range returns the Shomate's coefficients listed in Table 5.1. The thermophysical parameters of the substance are listed in Table 5.2

adopted, are consistent with the FPMD investigation of Baigain et al. (2015) (i.e., $K_0 = 24 \pm 2.5$ GPa at 3000 K; $K_1 = -0.0045$ GPa/K; $K_2 = 0$) has been assessed to deduce the isobaric heat capacity at high-T (Table 5.2). The molar volume of the fictive CaO liquid at room conditions was obtained by adopting a constant thermal expansion coefficient (i.e., $\alpha_V = 1.11 \times 10^{-4}$ K$^{-1}$), again consistent with the observations of Bajgain et al. (2015) at intermediate T (i.e., $\alpha_V = 1.167 \times 10^{-4}$ K$^{-1}$ at 3000 K).

### 5.2.8 CaO (Lime)

The vibrational properties of lime have been computed *ab-initio* at the B3LYP level of theory by Belmonte et al. (2017a). The phonon frequencies of three optic modes were obtained by vibrational analysis at Γ point of the Brillouin zone: one longitudinal optic mode (i.e. $\nu_6 = \nu_{LO} = 552.4$ cm$^{-1}$) and one (doubly-degenerate) transverse optic mode (i.e. $\nu_{4,5} = \nu_{TO} = 276.0$ cm$^{-1}$. Based on Born effective charges and dielectric tensor, the calculated LO-TO splitting was 276.4 cm$^{-1}$. This is in excellent agreement with neutron spectrometry (263.6 ± 23.8 cm$^{-1}$; Saunderson and Peckham, 1971). The elastic constant tensor analysis gave the acoustic frequencies at the Brillouin zone boundary ($\mathbf{k}_{max}$) ($\nu_1 = 140.5$ cm$^{-1}$; $\nu_2 = 141.6$ cm$^{-1}$ and $\nu_3 = 230.3$ cm$^{-1}$ [see

Ottonello et al. (2010b) and Belmonte et al. (2014, 2016), for full details about the calculation method].

The harmonic isochoric heat capacity conforms obviously strictly to the Dulong-Petit limit at high temperature and practically overlaps with the values obtained by Erba et al. (2015) through a full phonon dispersion calculation in the T range between 0 and 2000 K (Fig. 5.21). The isobaric heat capacity reproduces the experimental values of Robie and Hemingway (1995) in the intermediate T range (Fig. 5.21) when adopting selected thermophysical parameters from the literature. The Shomate's function for $C_P$ (Table 4.1) conforms to the expected path at high-T (i.e., $3nR + TV\alpha_V{}^2K_{(0,T)}$]. This is a compulsory requisite from a thermodynamic point of view, as stressed many times (e.g., Ottonello, 2024a, 2024b), but rarely fulfilled by literature assessments.

There are no experiments (to the best of our knowledge) concerning the high-pressure melting curve of lime. However, MD calculations predict a substantial increase in the melting temperatures with pressure (Sun et al., 2010). The thermal derivatives of the bulk modulus of solid CaO may be tuned in such a way to substantially reproduce the results of the MD simulation of Sun et al. (2010) up to pressures of about 70 GPa (Belmonte et al., 2017a; Fig. 5.22 and Table 4.2).

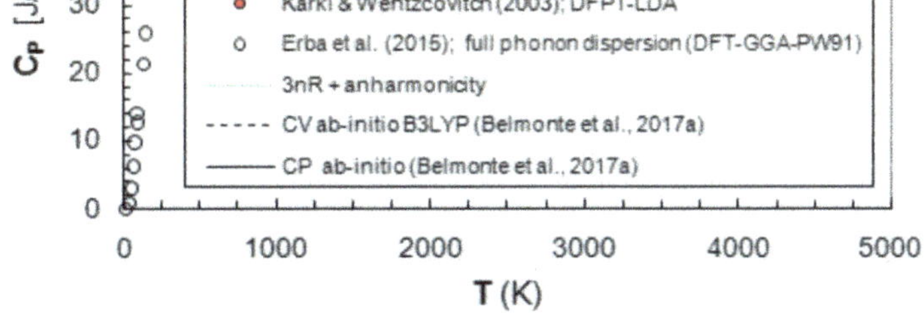


**Fig. 5.21** Ab-initio isobaric (and isochoric) heat capacity of solid CaO (lime) compared with the results of previous investigations and thermodynamic assessments. The red-dashed line is drawn only for illustrative purposes and shows the asymptotic behavior of the isobaric heat capacity induced by anharmonicity (i.e., $3nR + TV\alpha^2K$). Nonlinear minimization of the discretized values in the 298.15—5000 K T-range returns the Shomate's coefficients listed in Table 4.1. The thermophysical parameters of the substance are listed in Table 4.2

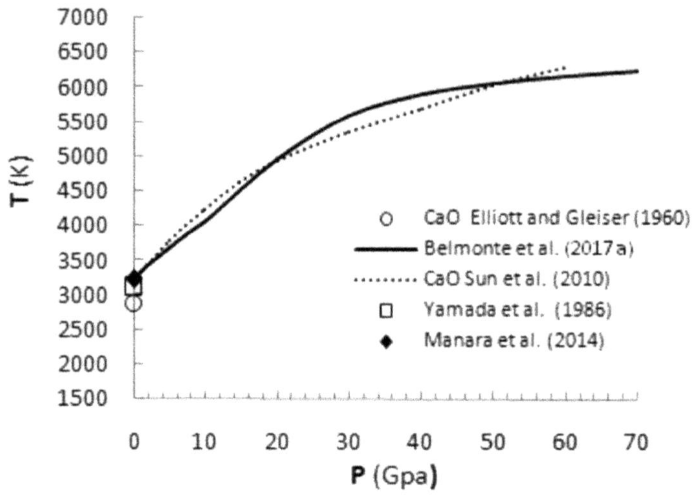


**Fig. 5.22** Melting curve of lime, compared with the MD results of Sun et al. (2010)

# Chapter 6
# Binary Joins

There are six binary joins in the CMAS quaternary system to be resolved before affording the four limiting ternary faces of the tetrahedron: (1) MgO-$SiO_2$; (2) CaO-$SiO_2$; (3) MgO-$Al_2O_3$; (4) CaO-$Al_2O_3$; (5) $Al_2O_3$-$SiO_2$; (6) CaO-MgO. The thermodynamic properties of mixtures in the liquid aggregation state of the first five joins are conveniently resolved in the framework of the Hybrid Polymeric Approach (HPA), while the 6$^{th}$ join may be simply modeled through a Redlich–Kister formalism adopting a low degree Van Laar expansion of interaction coefficients. HPA applications to the first principles investigation of the MgO-$SiO_2$ and CaO-$SiO_2$ joins have already been presented by us in the past (Ottonello 2024b; Belmonte et al. 2017a), and we must limit ourselves here to a short summary.

## 6.1 MgO-$SiO_2$

By applying:

$$\overline{H}_{mixing,T_f} = \frac{\left(H_{solid,T_f} + H_{fusion} - \sum n_i \mu^0_{i,liquid,T_f}\right)}{\sum\limits_i n_i} \tag{6.1}$$

One obtains indirect estimates of the enthalpy of mixing per unit mole of melt from experimental heats of fusion of the solids nucleating in the system. In the above equation, $n_i$ are the moles of oxide i per formula unit in the corresponding solid. Differential scanning calorimetry measurements by Richet et al. (1993) returned for $Mg_2SiO_4$ by a heat of fusion of 142 ± 14 kJ/mol at 2174 K. According to the NIST-JANAF tables (Chase 1998), the heat of fusion of $MgSiO_3$ is 75.3 ± 20.9 kJ/mol at 1850 K. Applying Eq. 6.1 in direct fashion, the expected enthalpy of mixing per unit

© The Author(s), under exclusive license to Springer Nature Switzerland AG 2026 
G. A. Ottonello et al., *Ab-initio Assisted Petrology*,
https://doi.org/10.1007/978-3-032-25658-4_6

mole of melt is $-18.5 \pm 4.7$ kJ/mol at $X_{SiO2} = 0.333$ and $-10.4 \pm 10.4$ kJ/mol at $X_{SiO2} = 0.5$. Applying the same equation in a reverse fashion returns an estimated enthalpy of fusion of 66.3 kJ/mol for $MgSiO_3$ and 154.7 kJ/mol for $Mg_2SiO_4$. Consider now that the polymerization constant of the liquid $K_P$ is expanded in the HPA model in the form

$$\mathrm{Ln}K_P = A/T + B \tag{6.2}$$

The A and B coefficients have a precise thermodynamic significance because R × A and −R × B represent, respectively, the enthalpic and entropic contributions to the chemical interaction energy between oxygen donors (NM) and oxygen acceptors (NF) per unit mole of NF-NM contacts in the melt.

Whenever there is a marked contrast in the donor–acceptor properties of the mixing oxides in the liquid state, the enthalpic interaction is dominant, and the enthalpy of mixing cannot be expected to vary appreciably with T. As we see in Fig. 6.1, the enthalpies of mixing generated through Eq. (6.1) are perfectly consistent (within experimental uncertainty) with the HPA model values.

In other terms, the first-principles procedure is substantially correct. This is a great advantage with respect to empirical parameterizations (which often give a correct appraisal in terms of Gibbs free energy of mixing but fail in deconvoluting the Gibbs free energy into H and S components).

Because in the case of interest here, there is a single NM component and a single NF on distinct (cationic and anionic) matrices, no ideal mixing contribution is present. The entropy of mixing is dominantly associated with the chemical

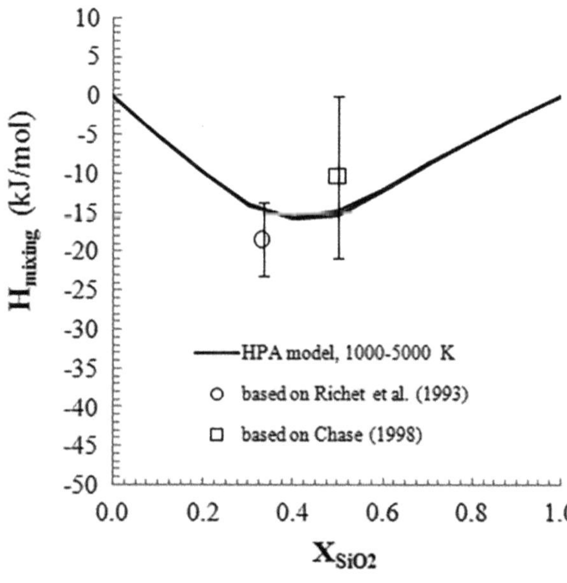


**Fig. 6.1** Enthalpy of mixing in the T range from 1000 to 5000 K. Estimates based respectively on heats of fusion of $Mg_2SiO_4$ and $MgSiO_3$ after Richet et al. (1993) and the NIST-JANAF tables (Chase 1998) are shown for comparative purposes. From Ottonello (2024b) with modifications

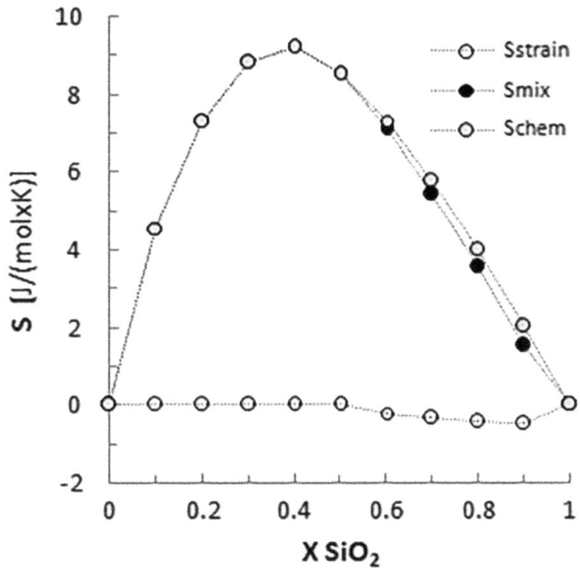


**Fig. 6.2** Entropy of mixing of the HPA model along the MgO-SiO$_2$ binary join. The entropy contributions associated with chemical interaction of the functionals (singly bonded oxygens) are dominant, while the terms associated with Hookian strain are quite limited

interaction and does not change appreciably with temperature. The strain entropy is largely subordinated (Fig. 6.2).

Although in the analysis of enthalpic and entropic components of the chemical and strain energies we benefit of the application of first principles, the assessment of volumetric terms is often entirely based on observations. These terms are in fact usually obtained from inverse methods based on "*topology*" (i.e., the experimental conformation of the liquidus curve at discrete P conditions. In this peculiar case (MgO–SiO$_2$ join), we benefit however from the few existing data based on First Principles Molecular Dynamics (FPMD) (de Koker et al. 2013) and also on MD with empirical potentials (Lacks et al. 2007). The calculations depict asymmetric contributions reproducible through a Van Laar model (Fig. 6.3).

To reproduce with accuracy the FPMD data, it is necessary to use a Redlich–Kister expansion of high degree (at least n = 2). However, because what is observed by petrologists (though of high quality) is often controversial, due perhaps to kinetic (nucleation ?) problems, it is more appropriate o adopt a low-order expansion (Ottonello 2024b). In fact, the 0$^{th}$ order parameter returns precisely the congruent melting point of the metasilicate (symmetric position of the compositional realm, unaffected by high-order parameters) and the 1$^{st}$ order parameter locates with sufficient accuracy the two invariant points at the opposite sides of the eutectic bulge.

The negative asymmetric contributions to V of the liquid mixture depicted by the HPA model are not excessively dissimilar to those obtained through FPMD calculations (see the first-order Redlich–Kister volumetric parameters at 5, 16, 24, and 26 GPa in Table A1 of Ottonello, 2024b and Eqn. 4.47). The baric evolution of the Van Laar model parameters requires a second-order polynomial expansion on P

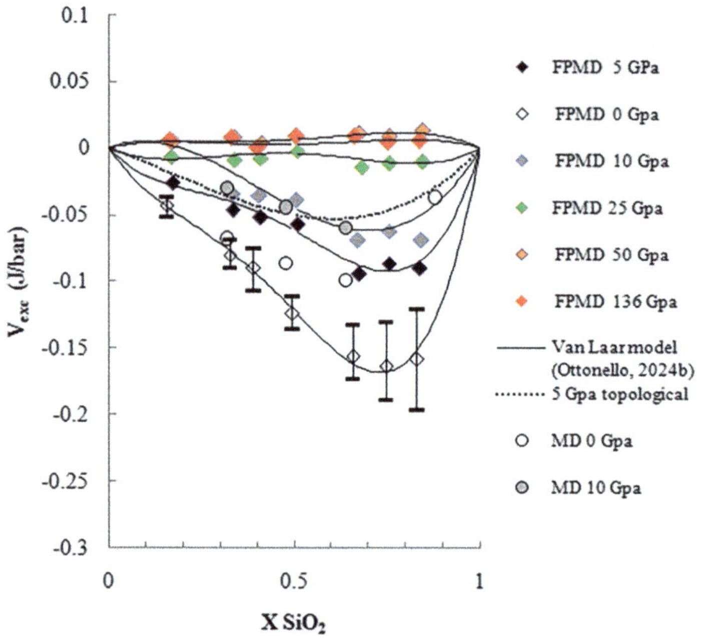


**Fig. 6.3** FPMD excess volumes (de Koker et al. 2013) at various P conditions compared with MD values (Lacks et al. 2007) and HPA topology-based computations at P = 5 GPa. The solid lines encompassing discrete FPMD values at constant P are Van Laar model computations based on a second-degree Redlich–Kister expansion. The dotted line encompasses Van Laar model computations based on first-degree Redlich–Kister expansions (topological reconstruction). From Ottonello (2024b) with modifications

(Fig. 6.4; Table 4.7). The $\eta_{0,V}$ of the HPA and FPMD procedures is affected in a similar manner by P. A direct comparison of $\eta_{1,V}$ terms is mathematically meaningless.

### *6.1.1 Intermediate Crystalline Compounds*

Disregarding the pure components MgO and $SiO_2$, only $MgSiO_3$ polymorphs (enstatite, clinoenstatite, protoenstatite, high-P clinoenstatite, majorite, ilmenite, perovskite), the $Mg_2SiO_4$ polymorphs (forsterite, wadsleite, ringwoodite), and the phase anhydrous B ($Mg_{14}Si_5O_{24}$) nucleate in the system (Tables 4.1 and 4.2).

All phases may be treated as pure compounds, contrary to what was anticipated by Schlaudt and Roy (1965) because the solubility of silica in MgO is negligible (Henriksen and Kingery 1979),

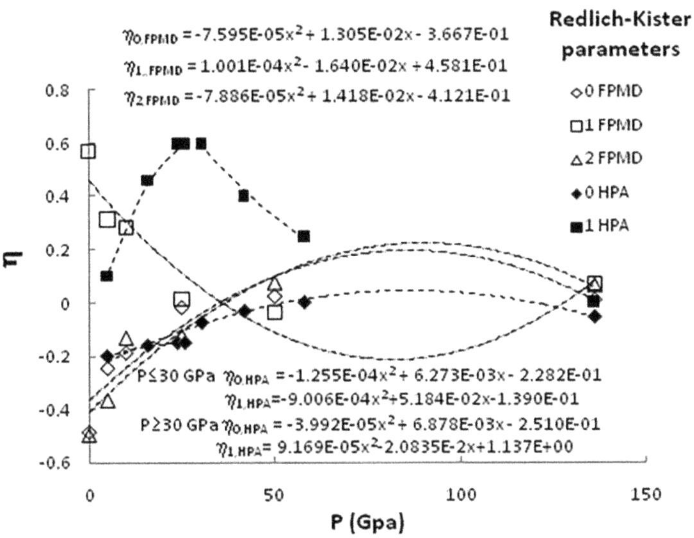


**Fig. 6.4** Baric evolution of Redlich–Kister parameters: the listed expansions are functions of P (in GPa). The second-order polynomial interpolations depict an apparent discontinuity at P $\approx$ 30 GPa

### 6.1.1.1 MgSiO$_3$ Polymorphs

MgSiO$_3$ crystallizes under various forms (*P2$_1$/c; C2c; PBca; Pbcn;* $R\overline{3}$;*I4$_1$/a; Pbnm*), in the realm of P–T conditions pertinent to our planet. This compound, being a major component of pyroxenes, has been the object of attentive investigation. Tables 4.1 and 4.2, respectively, list the thermodynamic and thermophysical properties adopted in this study.

#### 6.1.1.1.1 Orthoenstatite (Pbca)

Stangarone (2016) and Stangarone et al. (2016) investigated this phase ab-initio with the hybrid WC1LYP functional in the GGA approximation. A numerical integration over the unit cell volume returned the exchange–correlation contribution to the Fock matrix.

We adopt here a standard-state volume consistent with the value listed by Berman (1988) and based on the data of Chernosky et al. (1985) ($V^0_{298.15} = 31.34 \pm 0.02$ cc/mol), slightly lower than the relaxed volume at the athermal limit observed by Stangarone (2016) and Stangarone et al. (2016) ($V_0 = 31.90$ cc/mol; $K_0 = 90.91$ GPa; K' $K\prime = 10.86$).

Orthoenstatite has 240 optical modes (16 formula units per unit cell) and the symmetry assignment at Γ point is as follows (Stangarone 2016; Stangarone et al. 2016):

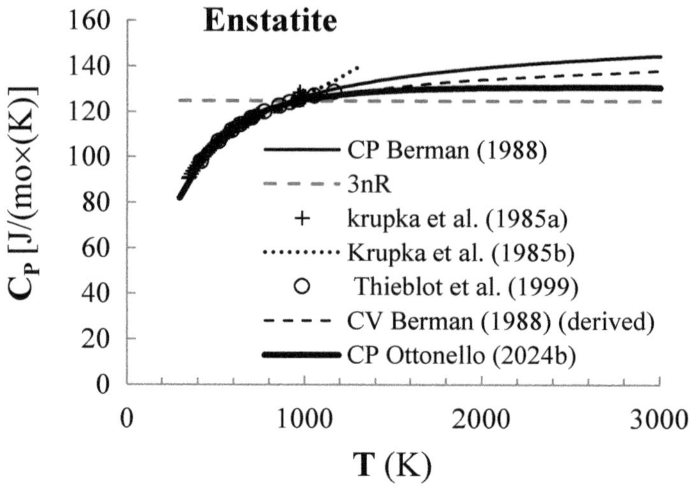


**Fig. 6.5** Isobaric ($P = 10^5$ Pa; solid lines) and isochoric dashed lines) heat capacity of *Pbca* Orthoenstatite (Ottonello 2024b; bold), compared with experiments (Krupka et al. 1985b; Thieblot et al. 1999) and assessment (Berman 1988; thin solid line). From Ottonello (2024b), with modifications

$$\Gamma = 30A_g + 30B_{1g} + 30B_{2g} + 30B_{3g} + 30A_u + 30B_{1u} + 30B_{2u} + 30B_{3u} \quad (6.3)$$

The $A_g$, $B_{1g}$, $B_{2g}$, and $B_{3g}$ modes are Raman-active (120). Modes $B_{1u}$, $B_{2u}$, and $B_{3u}$ are pure translations. The remaining (29 × 3) $B_{1u}$, $B_{2u}$, $B_{3u}$ are IR active while 30 $A_u$ modes are silent. Unfortunately, these silent modes have not been listed by Stangarone (2016) nor by Stangarone et al. (2016). The thermochemical computation of $C_V$ is thus impossible to carry out.

To circumvent this bias, we adopted thus as initial guess for $C_P$ the thermodynamic assessment of Berman (1988). It was, however, quite evident from the very beginning that Berman's $C_P = f(T)$ function (although reproducing the calorimetric experiments of Krupka et al. 1985b, and Thieblot et al. 1999) cannot be extended beyond the observational T range (the Dulong-Petit limit is virtually trespassed already below ~1200 K).

However, anharmonicity at various T and room pressure may be obtained adopting $V^0_{298.15}$ of Berman (1988), α from Skinner (1966), and $K_{(0,T)}$, $K'$, $K_1$,$K_2$ from Kung et al. (2004, 2005), Li and Liebermann (2007), and Hugh-Jones (1997) and the assessment of Saxena et al. (1993) [see Belmonte et al. (2017b) for details]. $H^0_{f,298.15}$ and $S^0_{298.15}$ are the refined values of Berman (1988) [based on the observations of Brousse et al. (1984) and Krupka et al. (1985b). The results of this exercise are shown in Table 4.1 and Fig. 6.5.

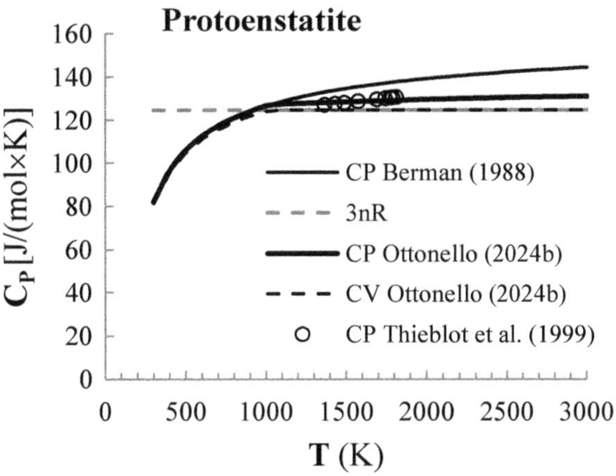


**Fig. 6.6** Isobaric ($P = 10^5$ Pa; solid line) and isochoric black dashed line) heat capacity of *Pbcn* Protoenstatite, compared with experiments (Thieblot et al. 1999). From Ottonello (2024b) with modifications

6.1.1.1.2 Protoenstatite (Pbcn)

This polymorph *has Pbcn* symmetry [Smyth and Ito (1977); see also Cameron and Papike (1981)]. *Ab-initio* investigations by Jahn and Martoňak (2009) at the DFT/LDA theory level with plane waves and optimized norm-conserving pseudopotentials gave a relaxed structure at the athermal limit with $a = 9.171$ Å, $b = 8.513$ Å, $c = 5.254$ Å, $V°_{298.15} = 30.88$ cc/mol. An *ab-initio* DFT/GGA investigation by Chibisov (2013) gave $a = 9.2947$ Å, $b = 8.8752$ Å, $c = 5.3553$, $V°_{298.15} = 33.24$ cc/mol, and $K_0 = 106$ GPa (presumably setting $K'_{(P,0)}$, to 4). It should be stressed, however, in passing, that the Birch-Murnaghan third-order bulk modulus was obtained in this study principally in the aim of clarifying the effect of P on the electronic properties, with particular emphasis on the bandgap.

Berman (1988) chose to assign to *Pbcn protoenstatite* the same $C_P$ of the *Pbca* polymorph *Orthoenstatite.* This expansion does not match the calorimetric data of Thiéblot et al. (1999) and trespasses the Dulong-Petit limit already at low-T (Fig. 6.6).

The $C_P$ obtained when subtracting the ($C_V$-3nR) heat capacity excess from the nominal one, however, matches satisfactorily the high-T calorimetric observations (Ottonello, 2024b; Fig. 6.6).

The intrinsic stability of pure protoenstatite is limited to ~1 GPa, hence its high-P-high-T properties are highly conjectural, and the adoption of linear $K_T = f(T)$ and $\alpha_T = f(T)$ is almost compulsory (Table 4.2).

Adopting $S^0_{298.15}$ of Berman (1988) and lowering his $H^0_{f,298.15}$ to ~350 J/(mol × K) the protoenstatite-Orthoenstatite boundary conforms to the observations of Presnall et al. (1998) (Fig. 6.7).

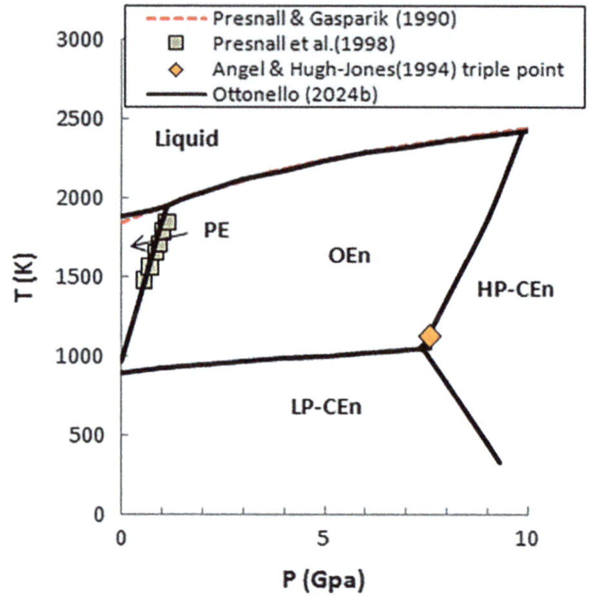


**Fig 6.7** Low-P solid-state transitions in $MgSiO_3$ polymorphs. Experiments and assessments are compared. From Ottonello (2024b) modified

6.1.1.1.3 *Low-P Clinoenstatite (P2₁/c)*

The first structural refinement is due to Kuno and Hess (1953) ($a$ = 9.618 Å, $b$ = 8.828 Å, $c$ = 5.186 Å, β = 71.38½°; Z = 4; monoclinic, space group $P2_1/c$,). The spatial assignment was confirmed by Morimoto et al. (1959) a few years later. An AIMD Full structural optimization of *Low-P Clinoenstatite* was performed by Duan et al. (2001), adopting the Damped Molecular Dynamics with Variable Cell Shape (VCS), an innovative procedure proposed by Wentzcovitch (1991), Wentzcovich and Martins (1991). Duan et al. (2001) adopted the Perdew-Zunger functional (Perdew and Zunger 1981) to parameterize the LDA exchange–correlation energy. The generated volumes gave $V°_{298.15}$ = 30.08 cc/mol, $K_0$ = 135 GPa, and K' = 4.9 (third-order Birch-Murnaghan fit). Due to the usual LDA over-binding, $V°_{298.15}$ is slightly underestimated with respect to experimental observations ($V°_{298.15}$ = 31.28 cc/mol; Angel and Hugh-Jones, 1994) and, for the same reason, $K_0$ is overestimated ($K_0$ = 111 GPa, and K' = 6.6 ± 1.1; Angel and Hugh-Jones 1994).

The calculations of Duan et al. (2001) were later repeated by Yu et al. (2010) at the DFT/LDA theory level, again with plane waves (and including a DFT/GGA procedure with the PBE functional). The application of density functional perturbation theory returned the phonon frequencies at Γ-point.

The cell parameters obtained by Yu et al. (2010) are in better agreement with experiments ($K_0$ = 121.4 GPa; K' = 5.5). The Quasi-Harmonic-Approximation (QHA) gave $\alpha = 2.5 \times 10^{-5}\ K^{-1}$ at T = 300 K. The QHA isobaric heat capacity at T = 298.15 was $C_P$ = 81.67 J/(mol × K). Obviously, being based on first principles, $C_V$ obeys the Dulong-Petit limit (Fig. 6.8).

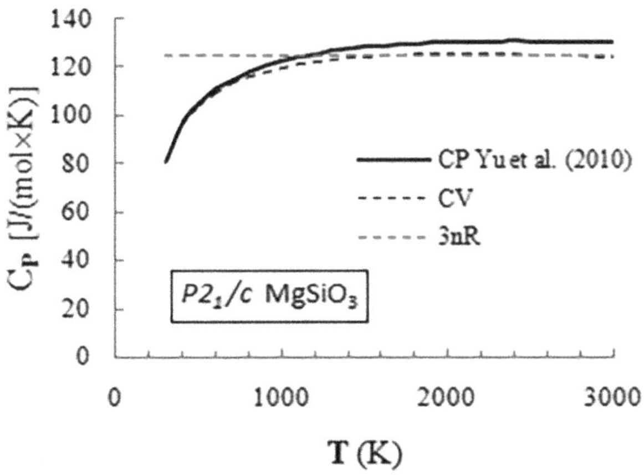


**Fig. 6.8** Isobaric (P = $10^5$ Pa; solid line) and isochoric (black dashed line) heat capacity of *P2₁/c* Low-P Clinoenstatite, based on the first principles investigation of Yu et al. (2010). From Ottonello (2024b) with modifications

In Table 4.1 are listed the Shomate expansion coefficients which reproduce the $C_P$ of Yu et al. (2010). The adopted α, $V^0_{298.15}$, $K_0$, K' are based on their investigation (Table 4.2). $K_1$ is from Saxena et al. (1993) (same Table). $H^0_{f,298.15}$) and ($S^0_{298.15}$) are from Ottonello (2024b) (Table 4.1).

The first principles parameterization reproduces within the experimental uncertainty the stability fields of the various low-P polymorphs (Fig. 6.7). We may also note in the same figure that the HPA liquidus of $MgSiO_3$ is virtually indistinguishable from the assessment of Presnall and Gasparik (1990).

### 6.1.1.1.4 High-P Clinoenstatite (C2/c)

*High-P Clinoenstatite* was investigated by Yu et al. (2010) with the same procedures previously seen for the low-P polymorph. Monoclinic, space group *C2/c* High-P, Low-P Clinoenstatite has a volume of 30.34 cc/mol, at T = 300 K (and a larger β angle with respect to *P2₁/c*), a bulk modulus of 126.3 GPa at the same T and K' = 5.5. The QHA thermal expansion obtained by Yu et al. (2010) at the same T conditions was $\alpha = 2.46 \times 10^{-5}$ $K^{-1}$ and $C_P$ was 81.1 J/(mol × K).

At DFT/GGA theory level (with the hybrid B3LYP functional) $V°_{298.15} = 31.808$ cc/mol, $K_0 = 101.6$ GPa, QHA $\alpha_{298.15} = 2.97 \times 10^{-5}$ $K^{-1}$ and QHA $C_{P,298.15} = 80.1$ J/(mol × K). (Belmonte 2013), The $C_P = f(T)$ function of the two *ab-initio* procedures are mutually consistent in a wide T range (Fig. 6.9). For the $H^0_{f,298.15}$) and ($S^0_{298.15}$) we adopt the values of Belmonte (2013) (Table 4.1). We adopt the results of Yu et al. (2010) for $C_P$, $V°_{298.15}$, $K_0$, K'(see Table 4.1 for the Shomate coefficients of $C_P$ and Table 4.2 for the α coefficients and the other thermophysical parameters). For $K_1$, we adopt the value indicated by Saxena et al. (1993) (Table 4.2).

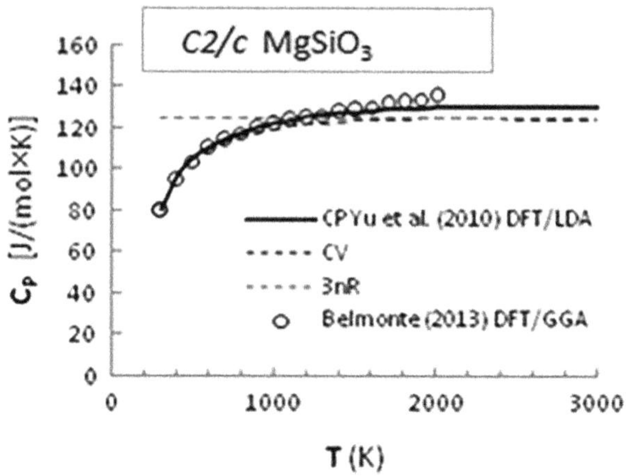


**Fig. 6.9** Isobaric (P = $10^5$ Pa; solid line) and isochoric (black dashed line) heat capacity of *C2/c* high-P Clinoenstatite, based on the first principles investigation of Yu et al. (2010) and Belmonte (2013). From Ottonello (2024b) with modifications

#### 6.1.1.1.5 Akimotoite-Majorite-Bridgmanite

*Akimotoite* is trigonal (ilmenite-type; space group $R\overline{3}$) with both Si and Mg octahedrally coordinated with O. Horiuchi et al. (1982) obtained by structural refinement $a = b = 4.7284$ Å; $c = 13.5591$ Å). $V°_{298.15}$ is thus 26.39 cc/mol.

*Majorite* is tetragonal (Space group *I4$_1$/a*) with an almost complete ordering of Mg and Si (origin of the non-cubic symmetry; Angel et al. 1989). A single-crystal XRD refinement of Majorite synthesized at 17 GPa and 1800 °C gave $a = b = 11.501$ Å, $c = 11.480$ Å, ($V_{cell} = 1518.6$ Å$^3$ → V = 28.58 cc/mol; Angel et al. 1989).

*Bridgmanite* is orthorombic (Space group *Pbnm*) with tilted $[SiO_6]^{8-}$ and distorted octahedra (the major differences from the ideal cubic perovskites). A single-crystal XRD refinement of Bridgmanite synthesized at 27 GPa and 1830 °C gave $a = 4.7754$, $b = 4.9292$, $c = 6.8969$ ($V_{cell} = 162.345$ Å$^3$, Z = 4 → V = 24.442 cc/mol; Horiuchi et al. 1987).

The three high-P polymorphs have been investigated *ab-initio* by Belmonte (2013), Hernández et al. (2015), and De la Pierre and Belmonte (2016). Hernández et al. (2015) employed in their DFT investigation a plane-wave basis set, combined with the Projector augmented Wave (PAW) method and the Perdew–Wang (PW91) functional. The relaxed Majorite structure of De la Pierre and Belmonte (2016) (DFT/GGA; same basis sets and computational parameters adopted for *C2/c* $MgSiO_3$) gave $a = b = 11.611$ Å, $c = 11.542$ Å, $V_{cell} = 1556.19$ Å$^3$ → $V_0 = 29.286$ cc/mol). Hernández et al. (2015) obtained for Majorite $V_{cell} = 1559.4$ Å$^3$ → $V_0 = 29.347$ cc/mol; $K_0 = 146.8$ GPa; K' = 4.6. For Akimotoite they obtained $V_{cell} = 270.65$ → $V_0 = 27.164$ cc/mol; $K_0 = 188.4$ GPa; K' = 4.6. For Bridgmanite they got $V_{cell} = 167.2$ → $V_0 = 25.173$ cc/mol; $K_0 = 229.7$ GPa; K' = 4.4.

We adopt in this study the parameters of Belmonte (2013) (Tables 4.1, 4.2) but we arbitrarily lower $H^0_{f,298.15}$ of the three polymorphs of −13 kJ/mol in order to make them stable at high P (predating in this way the hypothetical field of HP-CPx, which is limited to P < 14 GPa (Ottonello 2024b; Fig. 6.10).

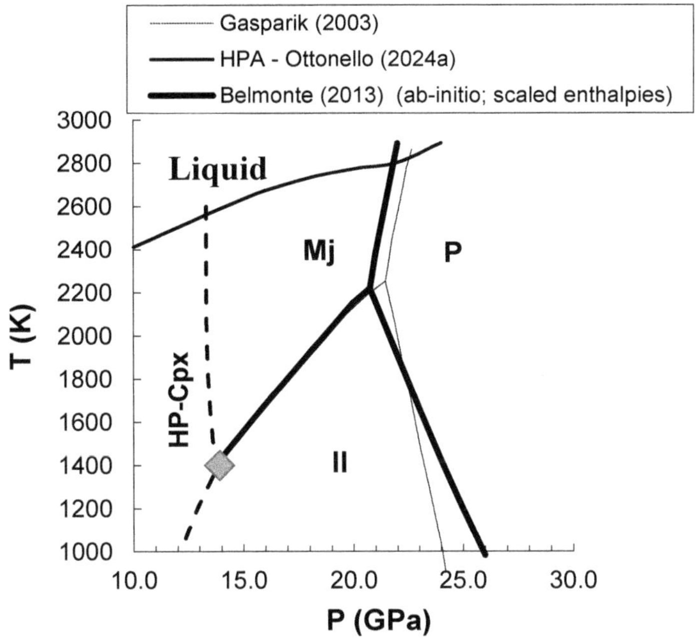


**Fig. 6.10** High-P solid-state transitions in $MgSiO_3$ polymorphs. Various assessments are compared with the results of the present study. The HPA liquidus is also shown. From Ottonello (2024b), with modifications

In sufficiently good agreement with the assessment of Gasparik (2003), the invariant triple point Mj-Pv-Ilm occurs at $P = 20.75$ GPa and $T \cong 2220$ K. This condition is intermediate with respect to other DFT-GGA and DFT-LDA assessments. Moreover, at $P \cong 13.9$ GPa and $T \cong 1400$ K a triple point appears, involving HP-CPx-Mj-Ilm.

As already seen for the low-P polymorphs (Fig. 6.7), the HPA approach seems appropriate in depicting the properties of the liquid even at these very high-P conditions.

#### 6.1.1.2 $Mg_2SiO_4$ Polymorphs

There are three $Mg_2SiO_4$ polymorphs. *Forsterite* (the α phase *Pbnm*), is stable below ~11 GPa and, at room pressure, melts congruently at $T = 2163 \pm 20$ K (Bowen and Andersen 1914). At $P \geq 11$ GPa *Wadsleyite* is stable (orthorhombic diphyramidal; *Imma*). The dP/dT slope of transition is slightly positive. The stability field of β-$Mg_2SiO_4$ is extrinsically limited at high P by the appearance of *ringwoodite* (γ-$Mg_2SiO_4$; $Fd\overline{3}m$). The slope of $\beta/\chi$ transition is sub-parallel to that of $\alpha/\beta$. Transition loci have been precisely identified by the experiments of Katsura and Ito (1989), Boehler and Chopelas (1991), and Fei and Bertkha (1999).

#### 6.1.1.2.1 Forsterite (Pbnm)

*Forsterite is* orthorombic bipyramidal (space group *Pbnm*; non-conventional setting of *Pnma* n. 62). Several first-principles investigations of forsterite have been provided through the years. These concerned both the elastic and structural properties at high pressure (Brodholt et al. 1996; Da Silva et al. 1997; Jochym et al. 2004), and the vibrational behavior (Noel et al. 2006; Li et al. 2007; Ottonello et al. 2009b).

The DFT/B3LYP GGA investigation of (Ottonello et al. 2009b) delineated a compressional behavior (*ab-initio* cell edges vs *ab-initio* static pressure) substantially consistent with the experimental evidence of Hazen (1976) at high P. The substance compresses anisotropically (preferentially along the *b* axis). The QHA $\alpha K$ product attains, at high-T, a quite constant value, averaging $4.47 \times 10^{-3}$ GPa/K in the T range $1000 \leq T \leq 3000$.

The QHA $C_P$ at room T is in substantial agreement with the calorimetric datum of Robie et al. (1978) [computed = 116.8 J/(mol × K), measured = 117.9 J/(mol × K)]. The agreement persists at high-T (see Tables 4.1, 4.2 for the adopted parameterization).

#### 6.1.1.2.2 Wadsleyite (Ibmm)

*Wadsleyite (*also known as "*beta form*" or "*modified spinel*") is *o*rthorombic bipyramidal (*Ibmm*). The B3LYP-GGA investigation of Ottonello et al. (2009b) gave results consistent with the results of Wu and Wentzcovitch (2007) obtained at the LDA level of theory. The $\gamma$-mode analysis of *ab-initio* frequencies yielded an average value of $\alpha K$ of $5.48 \times 10^{-3}$ GPa/K at high-T. This is in overall agreement with the experimental data of Meng et al. (1993) (an increase of the $\alpha K$ product $3.65 \times 10^{-3}$ GPa/K at 300 K to attain a constant value of $5.16 \times 10^{-3}$ GPa/K above its Debye temperature). The *ab-initio* thermal derivative of the bulk modulus was substantially smaller in magnitude than the experimental data of Fei et al. (1992) on $\beta$-[$(Mg_{0.84}Fe_{0.16})_2SiO_4$] (i.e., $K_1 = -0.012$ GPa/K against $-0.027$ GPa/K]. However, when coupling the thermal derivative of the bulk modulus with the $\alpha K$ product the resulting $\alpha$ is roughly consistent with the data of Suzuki et al. (1980) (Table 4.2).

The QHA low-temperature $C_P$ values of Ottonello et al. (2009b) are in good agreement with the calorimetric measurements of Akaogi et al. (2007), and those at high-T agree with the thermodynamic assessment of Jacobs and de Jong (2005) up to 1500 K. For higher T, the DFT/B3LYP computation gave values intermediate between those of Jacobs and de Jong (2005) and Watanabe (1982).

The standard-state B3LYP predicted entropy with scaled frequencies was $S°_{298.15} = 87.3$ J/(mol × K), which is intermediate between the value of $86.4 \pm 0.4$ J/(mol × K) determined by Akaogi et al. (2007), and the 88.7 J/(mol × K) predicted by Wu and Wentzcovitch (2007). Significantly higher values were proposed by Jacobs and de Jong (2005) [$S°_{298.15} = 91.3$ J/(mol × K)] and Saxena et al. (1993) [$S°_{298.15} = 93.0$ J/(mol × K)].

#### 6.1.1.2.3 *Ringwoodite* ($Fd\bar{3}m$)

*Ringwoodite* (the γ-polymorph) is cubic (space group $Fd\bar{3}m$). It is practically an *inverse spinel*, with Si occupying the tetrahedral site.

The obtained B3LYP-GGA volumes of Ottonello et al. (2009b) are intermediate between previous LDA (Haiber et al. 1997; Kiefer et al. 1999; Yu and Wentzcovich, 2006) and GGA (Piekarz et al. 2002) data. The zero point bulk modulus is $K_0 = 196.389$ GPa, to be compared with the extrapolated values of Meng et al. (1994) ($K_{(0,T)} = 182$ GPa), Weidner et al. (1984) ($K_{(0,T)} = 183$ GPa), and Jackson et al. (2000) ($K_{(0,T)} = 185$ GPa). The pressure derivative is K' = 4.322 (close to the value indicated by Meng et al. (1994), i.e., K' = 4.2. The thermal derivative is $K_1 = -0.010$ GPa/K, much lower with respect to the value indicated by Meng et al. (1993) ($K_1 = -0.028 \pm 3$ GPa/K). However, by coupling the *ab-initio* αK and $K_1$ one obtains a thermal expansion in close agreement with the experiments of Suzuki et al. (1979) and Ming et al. (1992).

The QHA $C_{P,298.15}$ of *ringwoodite* is in good agreement with the calorimetric data of Akaogi et al. (2007), while the high-T values are in good agreement with those of Chopelas et al. (1994), and Watanabe (1982).

The DFT/B3LYP third law entropy at 1 bar, 298 K (obtained with scaled frequencies) is $S°_{298.15} = 81.7$ J/(mol × K), close to the calorimetric value of $82.7 \pm 0.5$ J/mol-K (Akaogi et al. (2007) and significantly lower than the estimates of Fei et al. (1990) [$S°_{298.15} = 89.0$ J/(mol × K)], Jacobs and de Jong (2005) [$S°_{298.15} = 88.1$–$91.8$ J/(mol × K)] and Yu and Wentzcovitch (2006) [$S°_{298.15} = 85$ J/(mol × K)].

Figure 6.11 shows the extrinsic stability limits of the three polymorphs at subsolidus compared with experiments. The bias between *ab-initio* computations and direct experiments is around 2GPa for the $\alpha/\beta$ equilibrium. Scaled volumes (Table 4.2) restitute a Clapeyron slope consistent with observations at T,P of experiments. Raw *ab-initio* predictions for the $\beta/\gamma$ transitions are offset by roughly 4 GPa with respect to the most recent experiments (Fig. 6.11). Volume optimization restitutes a Clapeyron slope consistent with the experiments of Kuroda et al. (2000) and with the recent reversal experiments of Inoue et al. (2006).

### 6.1.1.3 $Mg_{14}Si_5O_{24}$ (Anhydrous Phase B)

This dense anhydrous silicate belongs to space group *Pmcb* (Finger et al. 1991). It was first synthesized by Herzberg and Gasparik (1989). Ganguly and Frost (2006) determined experimentally the equilibrium boundary of the reaction forsterite (Fo) + periclase (Per) = anhydrous phase B (Anh-B) at 900–1600 °C. Ottonello et al. (2010b) computed its thermodynamic-thermophysical properties in the framework of DFT/B3LYP under GGA approximation. The obtained static relaxed volume was $V°_{298.15} = 258.221$ cc/mol, the bulk modulus $K_0$ was 155.0 GPa, and its baric derivative K' = 4.14. The QHA analysis of the αK product returned a fairly constant value at high temperature, averaging 0.0050 GPa/K in the T range 1000–3000 K and attaining

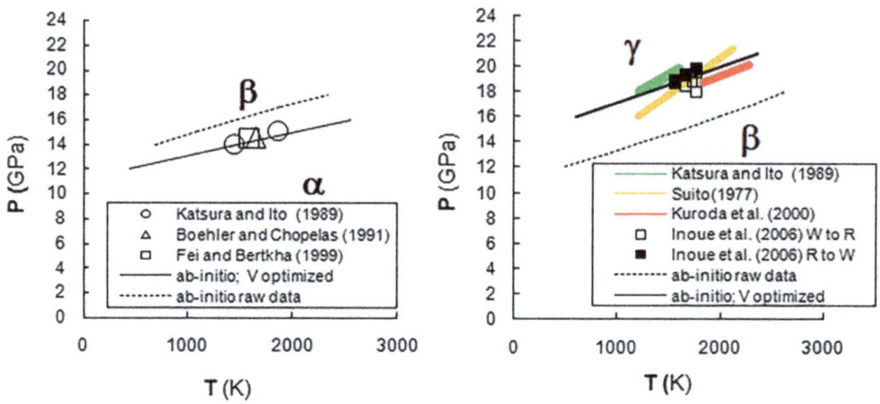


**Fig. 6.11** The univariant equilibrium between α and β polymorphs (left). and β and $\gamma$ (right). Results of all-electrons *ab-initio* calculations (Ottonello et al. 2009b) are compared with experimental observations. An overall agreement with experiments is obtained by setting the standard-state volumes of the α polymorphs to 4.367 J/bar (Matsui and Syono 1968) and scaling the β and γ polymorphs respectively to 4.0167 and 3.8933 J/bar. From Ottonello (2024b) with modifications

0.0051 GPa/K at the $T \rightarrow \infty$ limit. All thermophysical and thermodynamic parameters obtained in the quoted investigation are here retained, besides $H^0_{f,298}$, which was enhanced from −13,195.4 kJ/mol (Ottonello et al. 2010b) to −131,800.0 kJ/mol (still within the intrinsic precision of the *ab-initio* procedure) and the molar volume (we adopt here the datum of Finger et al. 1991). (Tables 4.1, 4.2). Recently, two new experiments on the extrinsic stability of Anh-B with respect to orthosilicate polymorph + periclase appeared (Koitani et al. 2017; Yuan et al. 2018). Figure 6.12 shows the results of the *ab-initio* investigation compared with experiments.

We stress that, while Ottonello et al. (2010b) adopted for forsterite and Periclase an ad-hoc selection of thermophysical-thermodynamic properties, we prefer here to rely completely on first principles for the various intervening phases, besides the standard-state molar volume of $Mg_2SiO_4$ polymorphs that were tuned on their own solid-state transitions (Fig. 6.10),

Based on our computation, the extrinsic stability field of Anh-B with respect to a $Mg_2SiO_4$ + MgO assemblage widens with the increase of T. Anh-B appears to predate the $Mg_2SiO_4$ + MgO assemblage in a P range comprised from roughly 9–23 GPa.

### 6.1.2 *The MgO-SiO$_2$ System at Room Pressure*

The conformation of the MgO-$SiO_2$ liquidus at room pressure has been investigated for more than one century. Forsterite was reported to melt congruently at T = 2161 K (Bowen and Andersen 1914) and to form an eutectic with periclase at T = 2136 K

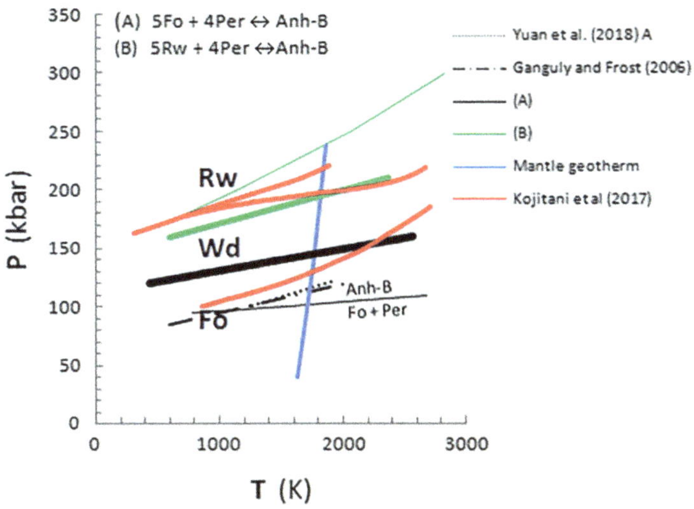


**Fig. 6.12** Extrinsic T-P stability field of Anh-B vs the $Mg_2SiO_4$ polymorphs + MgO assemblage. *Ab-initio* computations are compared with experimental evidence after Ganguly and Frost (2006), Koitani et al. (2017), Kojitani et al. (2017), and Yuan et al. (2018). At mantle geotherm conditions, Anh-B appears to be stable between roughly 9 and 23 GPa. The black and green heavy lines separate the stability fields of the various orthosilicate polymorphs. The mantle geotherm (blue solid line) is also plotted as an eye guide to the expected loci of transition. From Ottonello (2024b) with modifications

(Schlaudt and Roy 1965). A deep eutectic was observed between protoenstatite and cristobalite (T = 1816 K and $X_{SiO2} \cong 0.55$) near the protoenstatite peritectic (T = 1830 K, $X_{SiO2} \cong 0.51$). An extensive solvus opens at high $SiO_2$, from approximately $X_{SiO2} \cong 0.6$ to $X_{SiO2} \cong 0.98$ and a monotectic at T = 1967 K (Greig 1927). The consolute temperature is still the object of controversy (2263 K according to Hageman and Oonk 1986; 2473 K according to Ol'shanskii 1951). HPA reproduces the P = $10^5$ Pa liquidus satisfactorily within experimental uncertainty (Fig. 6.13).

Forsterite is predicted to melt congruently at T = 2166 K. The peritectic melting of protoenstatite occurs, according to HPA, at T = 1844 K, $X_{SiO2} \cong 0.52$. The eutectic between protoenstatite and β-cristobalite is observed at T = 1796 K, $X_{SiO2} \cong 0.58$. The solvus extends from $X_{SiO2} \cong 0.59$ to pure $X_{SiO2}$ and is caused by strain energy terms essentially entropic in nature (Table 4.7). The HPA consolute point is at T = 2244 K, $X_{SiO2} \cong 0.84$.

The miscibility gap at high $SiO_2$ content opens, due to the stabilizing effect of Network-Modifier cations (NM) on the σ-bonding Molecular Orbitals (MOs) of the oxygen framework (Gatti et al. 2012), and linear chains (Richet and Ottonello 2014). The ensuing increase in the force constant of Rigid Unit Modes (bending motions between adjacent monomeric groups) lowers the vibrational entropy of the substance. The substance remains metastably homogeneous until a critical instability region is attained. As schematically shown in the upper part of Fig. 6.14, the loci of intrinsic instability are depicted by unitary activity of the $SiO_2$ component in the

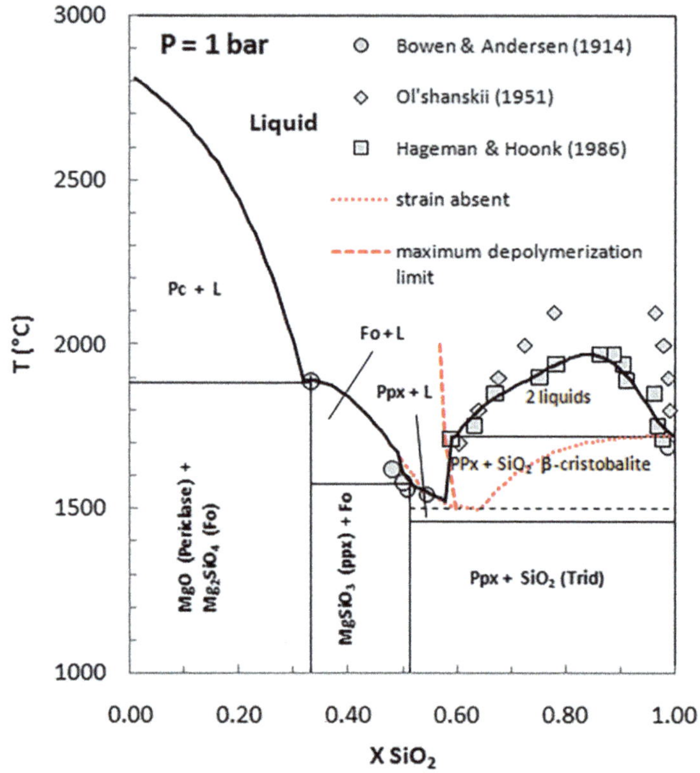


**Fig. 6.13** $MgO\text{-}SiO_2$. HPA model predictions at room pressure (heavy line) are compared with the experimental observations of Bowen and Andersen (1914), Ol'shanskii (1951), and Hageman and Hoonk (1986). The immiscibility occurs only in the presence of polymeric species. In the absence of strain energy (red-dotted line), immiscibility does not occur even at room pressure. From Ottonello (2024b) with modifications

liquid (solid state as reference). The silicate liquid indicates that this can be seen as a locally ordered aggregate of a $D_{6h}$ ($\alpha$-cristobalite) network and monomers in the short-medium range (Ottonello et al. 2010a). There is thus a close analogy with the solid state, and a slight increase in energy may thus favor nucleation and growth of crystal seeds.

The large biases between experimental sets are most likely due to different cooling rates and the degree of attainment of the intrinsic instability limit (lower part of Fig. 6.14).

The HPA model results in terms of computed activity, matching the room P experiments of Kambayashi and Kato (1984) (Fig. 6.15a), are satisfactory. The need to adopt a maximum depolymerization limit for the anion matrix is evident (Fig. 6.15b). along the guidelines anticipated long ago (Ottonello 2001). In fact, for a simple binary system such as $MgO\text{-}SiO_2$, the Temkin-like activity of the basic

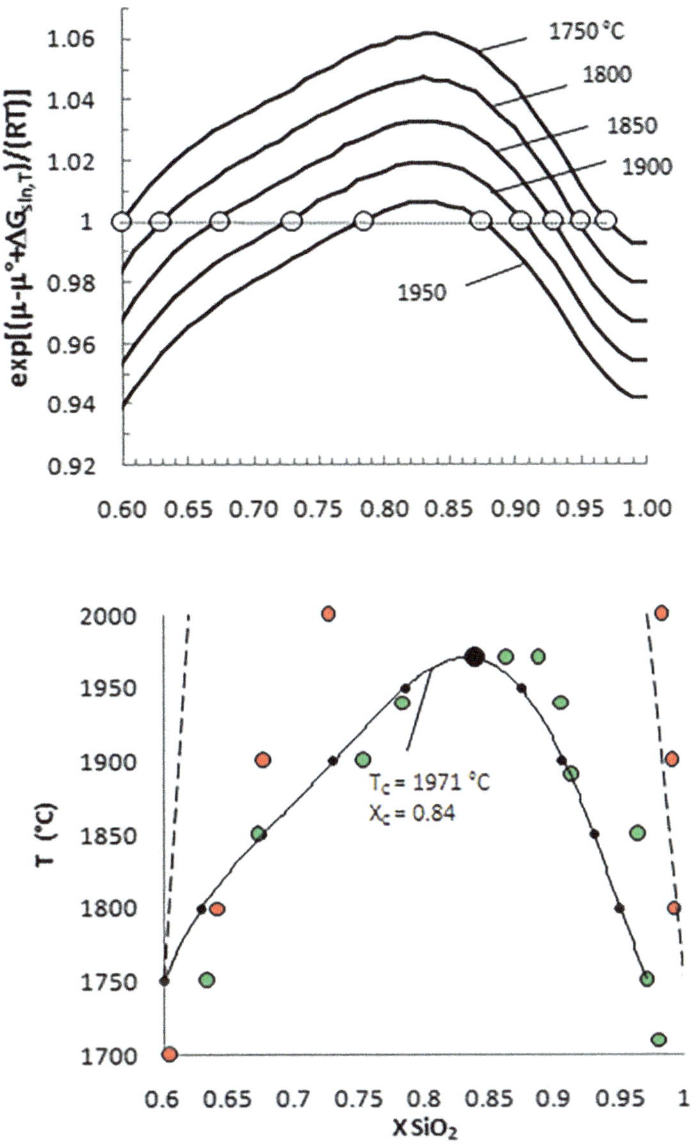


**Fig. 6.14** Intrinsic instability range depicted by the loci of unitary activity of $SiO_2$ in the liquid referred to solid at standard state (upper part of the figure) and projected onto the X-T space (lower part of the figure). The superposition of the experimental observations (Ol'shanskii 1951; red dots; and Hageman and Hoonk 1986; green dots) on the lower part of the figure explains the large biases between the two experimental sets as due to different cooling rates and degree of attainment of the intrinsic instability limit. From Ottonello (2024b) with modifications

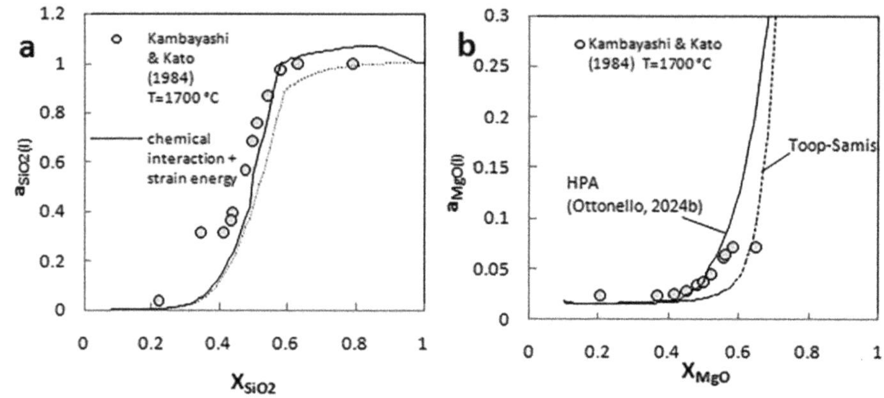


**Fig. 6.15** Thermodynamic activity of liquid components in mixture at T = 1700 °C compared with the experimental results of Kambayashi and Kato (1984), recast with pure liquid component as standard state of reference. From Ottonello (2024b) with modifications

oxide reduces to the molar fraction of free oxygen ions in the anion matrix, i.e. $a_{MO} = (O^{2-})/\sum \text{anions})$.

### 6.1.3 The $MgO$-$SiO_2$ System at High Pressure

The system was investigated through multi-anvil techniques the liquidus phase relations in the $0.5 \le X_{SiO2} \le 1.0$ sub-portion at P = 5 GPa. Protoenstatite was observed to melt congruently at 2235 K and coesite at 2753 K. A protoenstatite-coesite eutectic (68.8% in weight and 2203 K) was also detected. Unmixing is possibly still present at these P conditions (Dalton and Presnall 1997; Kanzaki 1990; Hudon et al. 2002), though sensibly reduced (Fig. 6.16). The negative deviations from Vegard's Law observed by de Koker et al. (2013) may be accounted for through a Redlich–Kister parametrization of the excess volume (Table 4.7). HPA predicts a congruent melting of Orthoenstatite at 2228 K. Forsterite reacts peritectically at $X_{SiO2} = 0.33$, T = 2380 K, and forms an eutectic with Orthoenstatite at $X_{SiO2} = 0.46$, T = 2211 K. The eutectic between Orthoenstatite and coesite is predicted to occur at $X_{SiO2} = 0.57$ and T = 2186 K.

Liebske and Frost (2012) carried out multi-anvil experiments at higher pressure (16 to 26 GPa) in the $0.0 \le X_{SiO2} \le 0.5$ sub-portion of the system. They observed at P = 16 GPa a peritectic reaction of the liquid with periclase at T ≈ 2603 K and $X_{SiO2} \approx 0.43$, leading to the formation of Anh-B. The same phase was shown to react peritectically at T ≈ 2560 to yield Wadsleyite. The same phase formed apparently an eutectic with Majorite at ~2510 K and $X_{SiO2} \cong 0.44$ (see Fig. 3.1a in Liebske and Frost (2012).

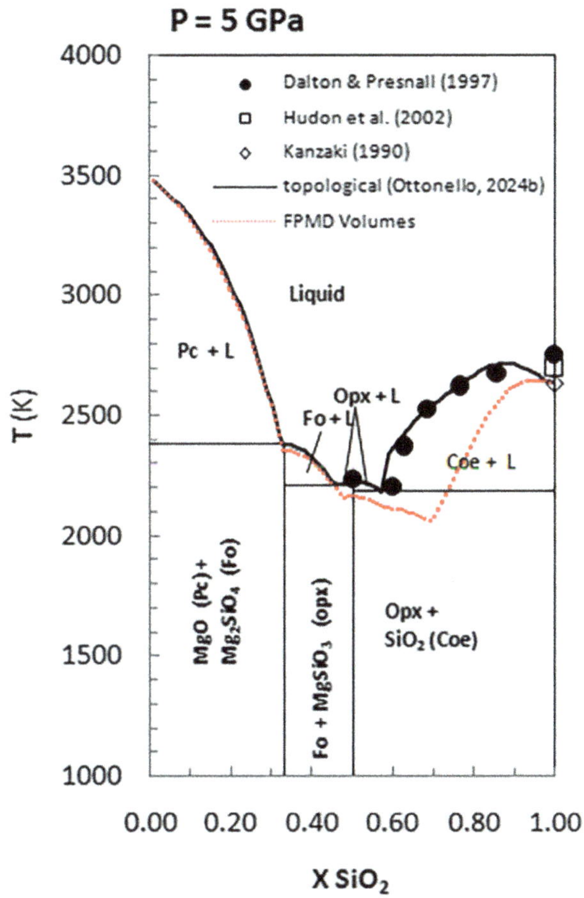


**Fig. 6.16** $MgO-SiO_2$. HPA model predictions at P = 5 GPa (heavy line) are compared with the experimental observations of Dalton and Presnall (1997), Hudon et al. (2002), and Kanzaki (1990). Congruent melting of orthopyroxene is predicted to occur at 2228 K. Forsterite is predicted to form by peritectic reaction at $X_{SiO2} = 0.33$ and T = 2380 K and to form an eutectic with Orthoenstatite at $X_{SiO2} = 0.46$ and T = 2211 K. The eutectic between Orthoenstatite and coesite is predicted to occur at $X_{SiO2} = 0.57$ and T = 2186 K

HPA calculations (Ottonello 2024b; Fig. 6.17) return the peritectic formation of Anh-B at the expense of Periclase at T = 2597 K and $X_{SiO2} \cong 0.40$. Two eutectics are seen to form: one between Anh-B and Majorite at 2503 K and $X_{SiO2} \cong 0.43$, and a second one between Majorite and Stishovite at 2542 K and $X_{SiO2} \cong 0.54$. Majorite is seen to melt congruently at T = 2561 K.

There is no HPA evidence of peritectic formation of Wadsleyite, while the liquidus with Majorite is correctly reproduced. As shown in Fig. 6.17, hindering the formation of Anh-B, the first eutectic would occur at $X_{SiO2} \cong 0.42$ and T = 2487 K and would involve Periclase (red-dashed line in Fig. 6.17).

According to the multi-anvil experiments of Liebske and Frost (2012) carried out at P = 26 GPa, Mg-perovskite is the stable polymorph of $MgSiO_3$ at liquidus and forms an eutectic with Periclase at $X_{SiO2} = 0.43$ and T $\cong$ 2800 K.

According to HPA calculations (Ottonello 2024b; Fig. 6.18), periclase forms an eutectic with Mg-perovskite at 2999 K and $X_{SiO2} = 0.46$. Perovskite is seen to melt

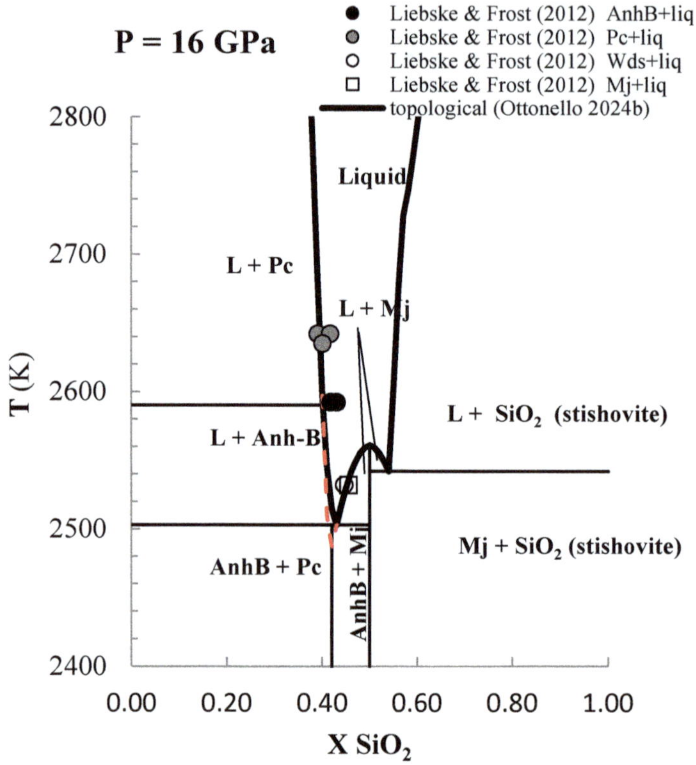


**Fig. 6.17** MgO-SiO$_2$. HPA model predictions at P = 16 GPa (heavy line) are compared with the experimental observations of Liebske and Frost (2012). Periclase reacts peritectically to form Anh-B at T = 2590 K and $X_{SiO2}$ = 0.42. Anh-B forms an eutectic with Majorite at 2542 K and $X_{SiO2} \cong 0.43$. Majorite forms an eutectic with Stishovite at T = 2596 and $X_{SiO2}$ = 0.54 and melts congruently at T = 2615 K. Preventing the formation of Anh-B, the first eutectic is between Periclase and Majorite and occurs few degrees below the true one (red-dashed line). Using interpolated Redlich–Kister coefficients ($\eta_{0,V} = -0.16$; $\eta_{1,V} = 0.46$) instead of the optimized ones ($\eta_{0,V} = -0.15$; $\eta_{1,V}$ = 0.43; Table 4.7) depresses somewhat the liquidus (dotted line). From Ottonello (2024b) with modifications

congruently at T = 3016 K and to form an eutectic with Stishovite at T = 2998 K and $X_{SiO2}$ = 0.54.

Quite recently, Yao et al. (2021) carried out experiments at P = 24 GPa, T = 2650 ± 100 K and T = 2750 ± 100 K (same P) on the SiO$_2$-rich side of the join ($Mg_4Si_6O_{16}$ and $Mg_3Si_7O_{17}$). The electron microprobe analysis of the quenched melt defined an eutectic melting between Perovskite and Sishovite at T = 2750 ± 100 K; $X_{SiO2}$ = 0.53 ± 0.03).

HPA calculations predict Perovskite to melt congruently at T = 2893 K. An eutectic between Periclase and Perovskite is observed at T = 2875 K and $X_{SiO2}$ = 0.46, and another one between Perovskite and Stishovite is observed at T = 2882 and $X_{SiO2}$ = 0.53 (Fig. 6.19). The excess volume parameters of the liquid that

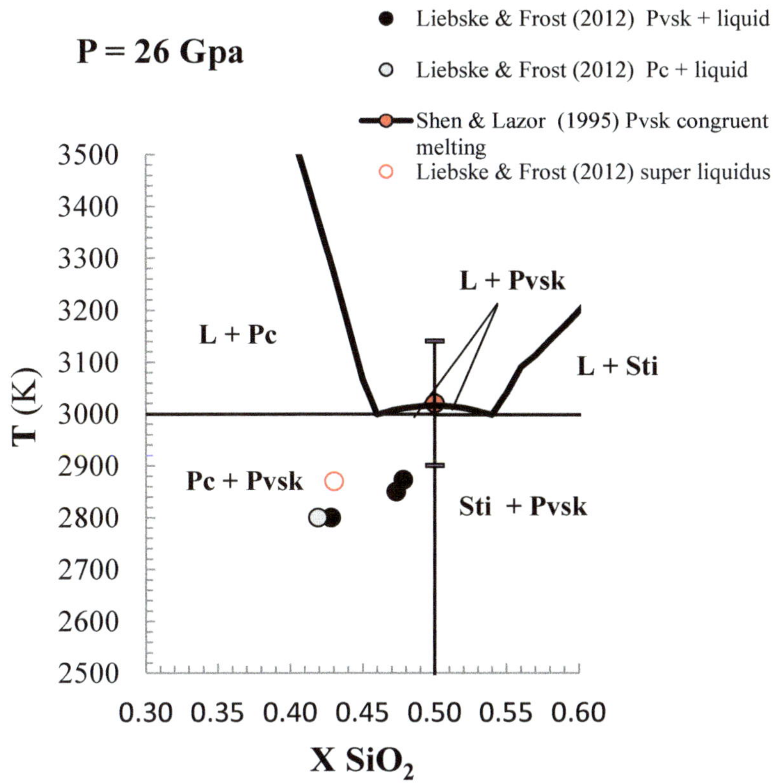


**Fig. 6.18** $MgO\text{-}SiO_2$. HPA model predictions at P = 26 GPa (heavy line) are compared with the experimental observations of Liebske and Frost (2012). Periclase forms an eutectic with Mg-Perovskite at 2999 K and $X_{SiO2} = 0.46$. Perovskite melts congruently at T = 3016 K and is seen to form an eutectic with Stishovite at 298 K and $X_{SiO2} = 0.54$

best conform to the (somewhat contrasting) topological evidence at the investigated pressures are reported in Table A1 of Ottonello (2024b).

The attainment of reasonable results for intermediate compositions such as $MgSiO_3$ is indicative of model reliability. The bias between topology-based assessments and experiments is, in some cases, dramatic (see Fig. 6.20), and the experimental observations of Zerr and Boehler (1993), Shen and Lazor (1995), and the FPMD computations of de Koker et al. (2013) are particularly apt to achieve a reasonable assessment. By selecting a few points along the $MgSiO_3$ melting curve, one may depict a possible baric evolution of the $MgO\text{-}SiO_2$ liquidus in the Earth's Mantle. The results of this exercise are summarized in Fig. 6.21.

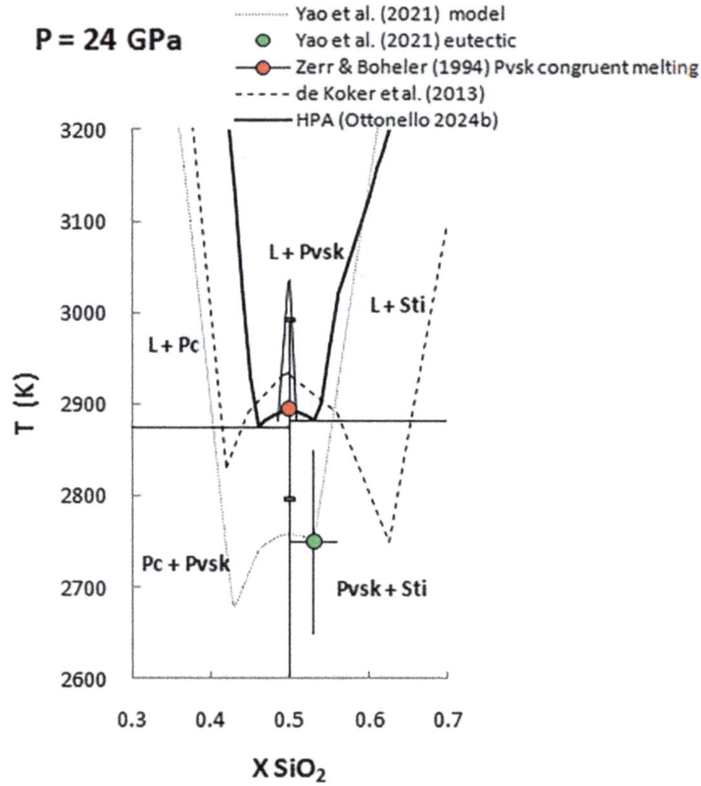


**Fig. 6.19** MgO-$SiO_2$. HPA model predictions at P = 24 GPa (heavy line) are compared with the experimental observations of Yao et al. (2021). The eutectic between Periclase and Mg-Perovskite and Stishovite is predicted to occur at 2875 K and $X_{SiO2}$ = 0.46, and the eutectic between Mg-Perovskite and Stishovite is seen to occur at T = 2882 and $X_{SiO2} \cong 0.53$. Perovskite is predicted to melt congruently at T = 2893 K

## 6.2 CaO-$SiO_2$

As anticipated, HPA applications to the first principles investigation of the CaO-$SiO_2$ joins have already been presented by us in the past (Belmonte et al. 2017a), and we must limit ourselves here to a short summary of what was previously observed (with limited readjustments).

From the topological analysis of the phase diagram, it turns out that the chemical interaction is completely enthalpic (Table 4.7; B = 0). The heat of mixing in the liquid, arising from the A model parameter, may be compared with the value obtained from the latent heat of fusion of solid compounds in the system, applying Eq. (6.1), as previously done for the MgO-$SiO_2$ binary join. We may thus see that the heat of

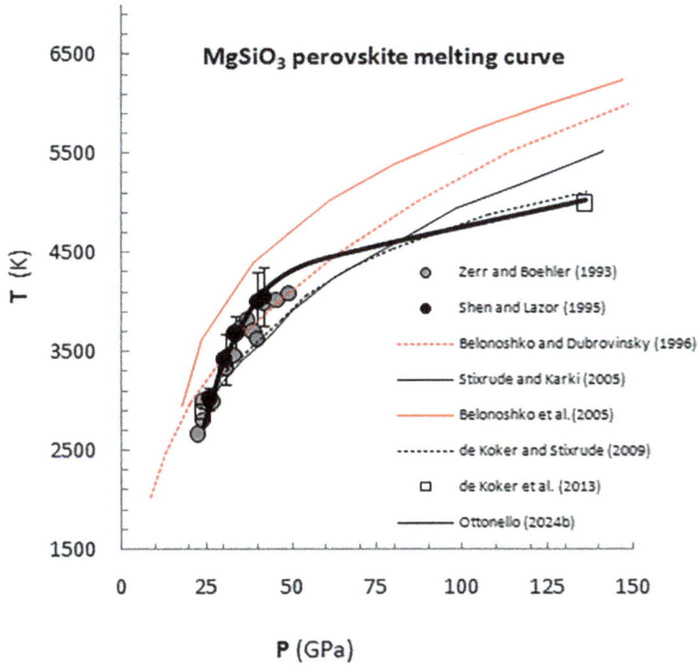


**Fig. 6.20** Melting curve of the Perovskite polymorph of metalisicate $MgSiO_3$ in a wide pressure range, compared with experiments and results of semi-empirical and first-principles MD calculations. The solid line encompasses the few points here adopted to conform to the P-dependent excess terms of the HPA model (see Table A1 in Ottonello 2024b)

mixing is quite consistent with the latent heats of fusion of the ortho- and meta-silicates nucleating in the system (α-$Ca_2SiO_4$ and pseudo-wollastonite; Barin 1995; Adamkovicova et al. 1980) (Fig. 6.22).

### *6.2.1 Intermediate Crystalline Compounds*

Besides lime and the various forms of silica, other twelve compounds are known to nucleate in the system: hatrurite ($Ca_3SiO_5$), five polymorphs of the orthosilicate $Ca_2SiO_4$) [$\alpha'_H$, $\alpha'_L$, α (bredigite), β (larnite), γ (Ca-olivine)], rankinite ($Ca_3Si_2O_7$), Ca-titanite ($CaSi_2O_5$)] and five polymorphs of the metasilicate $CaSiO_3$ (wollastonite, pseudo-wollastonite, walstromite, Wollastonite-II, Ca-perovskite). The extrinsic stability relations of the various polymorphs at subsolidus in the wide P,T regimes of interest in petrological investigations are quite complex and affect as well the melting behavior in a complex manner.

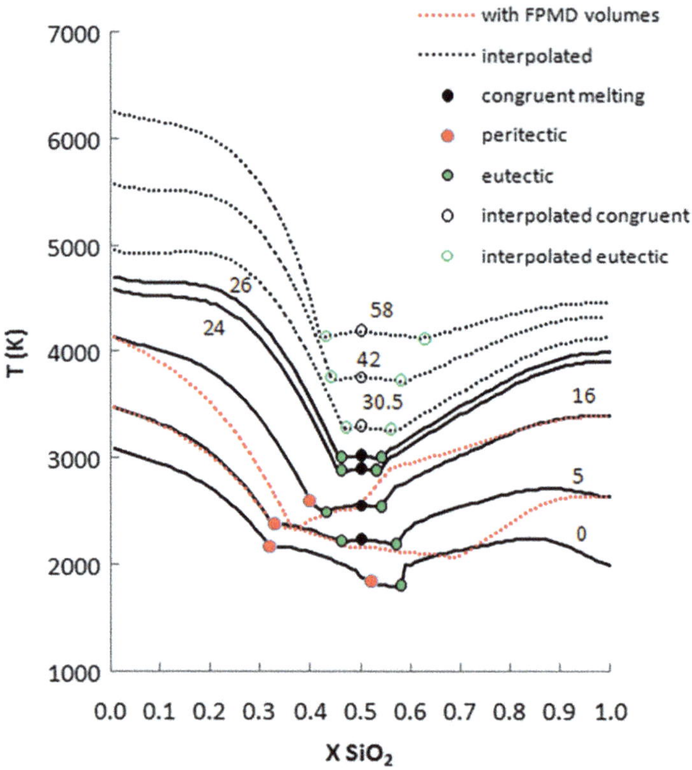


**Fig. 6.21** The MgO-SiO$_2$ liquidus at discrete P sampled along the MgSiO$_3$ melting curve. The metasilicate melts peritectically at room pressure, then congruently at higher P. From Ottonello (2024b) with modifications

#### 6.2.1.1 *Ca₂SiO₄*

At low-T and room P, the stable form is orthorombic (γ *Ca-olivine*), readily replaced by a monoclinic form (*β-larnite*) with the increase of P. The $\alpha'_H$ and $\alpha'_L$,, orthorombic forms stable at intermediate T (from ~1000 K to ~1700 K) transform into the α (*Bredigite*) polymorph, whose structural group (trigonal or exagonal) is still uncertain (see Remy et al. 1997 and references therein).

Bredrigite melts at room pressure at 2403 K. A complex extrinsic stability scheme has been proposed by Niesel (1972) to whom we refer the interested reader for reference.

As the common *lattice* isochoric heat capacity of β, α′, α, γ polymorphs Belmonte et al. (2017a) adopted the simple two-parameter expansion proposed by Berman and Brown (1985) for the $\alpha'$ polymorph (see Table 4.1). This expansion conforms rather well even to the extreme T relevant for the inner part of our planet up to the mantle/core boundary). Figure 6.23 shows the performance of this simple expression when

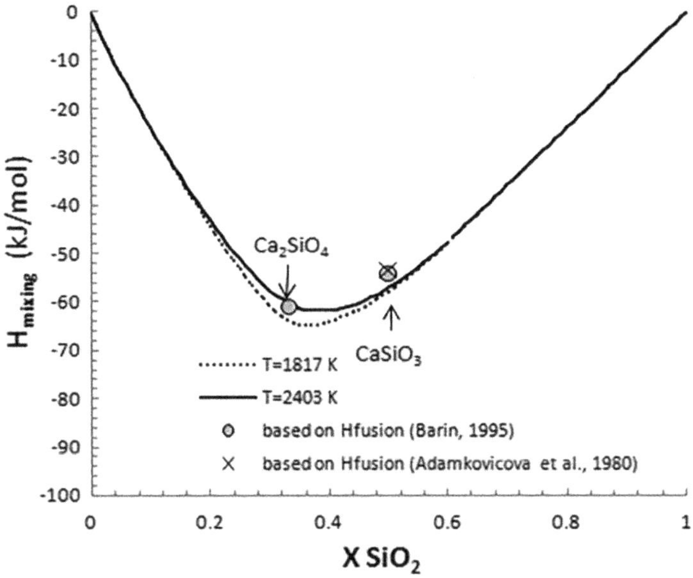


**Fig. 6.22** HPA heat of mixing along the CaO-$SiO_2$ join based on liquidus topology (black lines). The values expected from the latent heats of fusion of the two pure compounds at $T_f$ (Barin 1995; Adamkovicova et al. 1980) are superimposed for comparison. From Belmonte et al. 2017a, with modifications

compared with the low-T calorimetric data of Coughlin and O'Brien (1957) on *β-larnite*.

The $\beta - \alpha'$ transition was modeled by Berman and Brown (1985) as pure first order. The $\alpha' - \alpha$ transition has superimposed λ and first-order characters, with a sensible first-order contribution. We retain here their estimates and adopt their procedure to depict the heat capacity in the transition zone (Fig. 6.24; Table 4.3).

We disregard here the $\alpha'_H - \alpha'_L$ transition and treat $\alpha'$ as a single phase (the heat capacity functions of the α and γ polymorphs are shown in Fig. 6.25).

The adopted $H^\circ_f$ and S° for all $Ca_2SiO_4$ polymorphs are those of Belmonte et al. (2017a) (a readjustment of the parameters of Haas et al. 1981; carried out in order to obey the observed extrinsic stability limits).

The thermophysical properties of the various $Ca_2SiO_4$ polymorphs (Table 4.2) are essentially based on the observations of Remy et al. (1997) with some adjustments to conform to the attainment of the Dulong-Petit asymptotic limit at extreme T (i.e., the red-dotted lines in Figs. 6.23 and 6.25). Belmonte et al. (2017a) noted that the existing data about the thermal expansion of the various solid polymorphs of $Ca_2SiO_4$ (Saalfeld 1975; Barnes et al. 1980; Forest 1971; Remy et al. 1997, and references therein) are quite scattered and proposed the adoption of a "*generalized*" thermal expansion (Fig. 6.26).

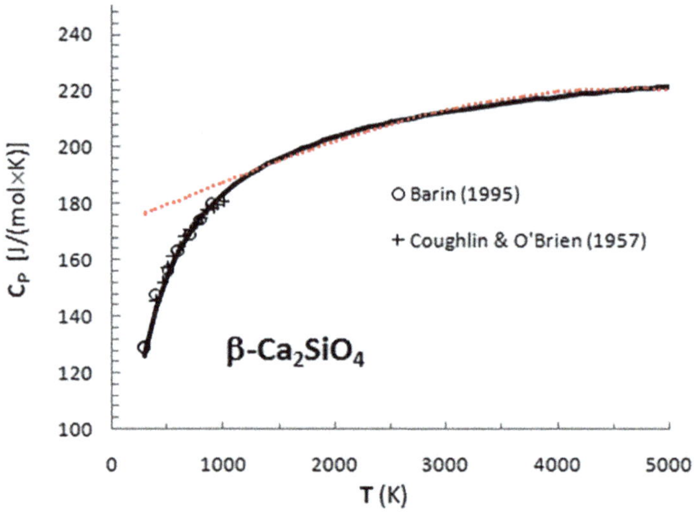


**Fig. 6.23** Isobaric heat capacity of β-larnite. The extended Shomate equation coefficients here adopted (Table 4.1) are the $C_P$ parameters proposed by Berman and Brown (1985) for the $\alpha'$ polymorph (i.e., a constant term and an inverse square root dependence on T). They are consistent with the low-T calorimetric data of Coughlin and O'Brien (1957) and Todd (1951) (not shown) and conform satisfactorily to the Dulong-Petit asymptotic limit (i.e., $3nR + TV\alpha^2K$; red-dotted line). The more recent tabulation of Barin (1995) is superimposed for comparative purposes. No refitting of the heat capacity function was deemed necessary

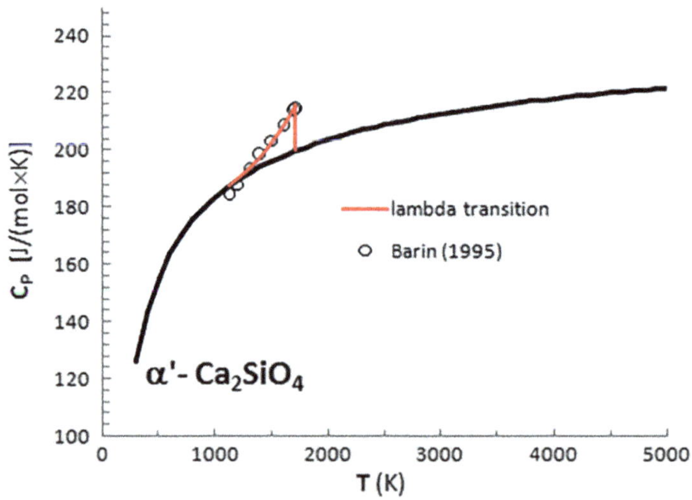


**Fig. 6.24** Isobaric heat capacity of the α' polymorph. The black continuous line is the lattice contribution depicted by the extended Shomate equation coefficients here adopted (Table 4.1). The continuous red line in the transition zone is associated with λ contributions $l_1$ and $l_2$ (Eq. 4.2a in Berman 1988; Table 4.3). The tabulation of Barin (1995) is superimposed for comparative purposes

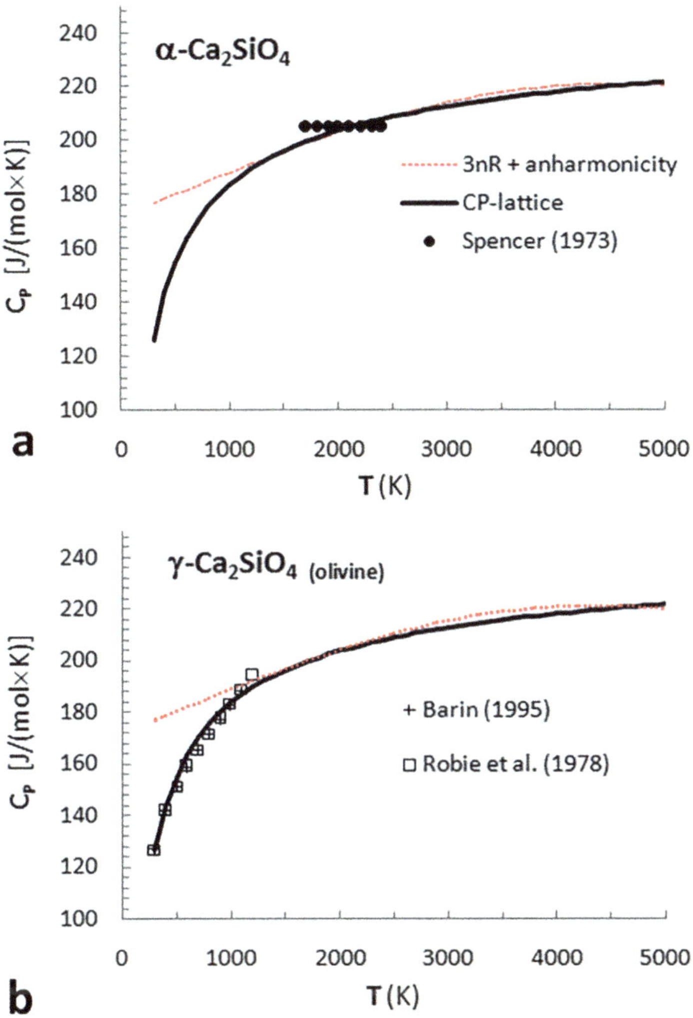


**Fig. 6.25** Isobaric heat capacity of the α and γ polymorphs. The estimates of Spencer (1973) for the α polymorph are superimposed for comparative purposes in part **a** of figure, and the tabulations of Robie et al. (1978) and Barin (1995) for the γ polymorph are shown in part **b**. The solid line is, in both cases, the simple 2-parameter "*lattice*" Cp equation

After this quite crude selection, they superimposed the thermally expanded volumes to the discrete observations of Remy et al. (1997) and selected the opportune standard-state volumes (Fig. 6.27 and Table 4.2).

The results of the computational exercise may be seen in the subsolidus phase diagram of $Ca_2SiO_4$ (Fig. 6.28).

The γ/β transition involves a negative Clapeyron slope. The $\alpha/\alpha'$, $\alpha'/\beta$ and $\gamma/\beta$ univariant curves have positive Clapeyron slopes and are consistent with the

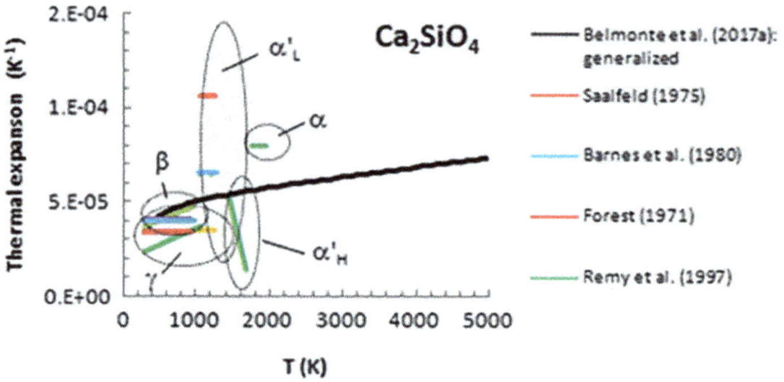


**Fig. 6.26** Thermal expansion of $Ca_2SiO_4$ polymorphs. Modified from Belmonte et al. (2017a)

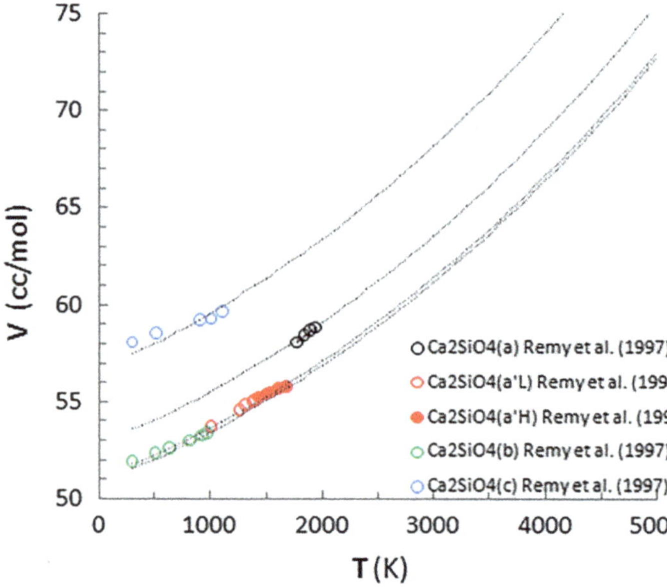


**Fig. 6.27** Thermally expanded volumes of the various $Ca_2SiO_4$ polymorphs. The modeled volumes (Belmonte et al. 2017a; dotted lines) are compared to discrete experimental observations after Remy et al. (1997). Modified from Belmonte et al. (2017a)

anticipations of Remy et al. (1995). Moreover, the $\alpha' / \beta$ univariant curve occurs not far from the phase transition detected by Wang and Weidner (1994) at ~9 GPa and 1350 K. $\alpha'$ is the stable polymorph at liquidus above ~5 GPa. The red solid line is the liquidus depicted by HPA. A nearly similar diagram (though limited to subsolidus) was recently proposed by Xiong et al. (2016).

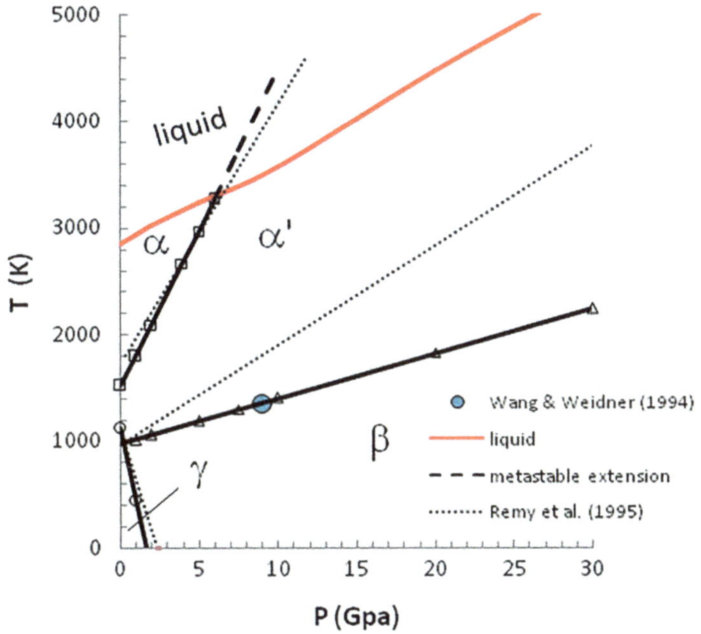


**Fig. 6.28** Computed phase equilibria in the system $Ca_2SiO_4$. The symbols are the results of discrete computations. The solid lines through them imply almost constant Klapeyron slopes consistent with constant ΔV/ΔS. The $\alpha / \alpha'$, $\alpha' / \beta$ and $\gamma / \beta$ univariant curves are consistent with the anticipations of Remy et al. (1995) and not far from the $\alpha'/\beta$ phase transition detected by Wang and Weidner (1994) at ~9 GPa and 1350 K. α' is the stable polymorph at liquidus above ~5 GPa. A nearly similar diagram was recently proposed by Xiong et al. (2016). From Belmonte et al. (2017a) with modifications

#### 6.2.1.2 $CaSiO_3$

The polymorphism of $CaSiO_3$ has been the object of detailed experimental investigations. At room conditions, *Wollastonite* (Triclinic, space group $P\overline{1}$ No. 2) is stable. According to Richet et al. (1991), there is at 995 ± 10 K a reversible transition from *Low-Wollastonite* to *High-Wollastonite.* The transition "*can be considered calorimetrically as first order*" (Richet et al. 1991).

Wollastonite (Wo) exhibits a solid-state transition to the Cyclosilicate *Pseudo-Wollastonite* (PsWo). According to Kushiro (1964), at room pressure, the phase transition takes place at $T_t = 1398 \pm 10$ K. According to Osborn (1942), PsWo melts congruently at 1817 K. Transposition on the 1968 T-scale places the T of fusion at 1821 K (Richet et al. 1991).

At high pressure (i.e., above roughly 2.5 GPa), PsWo transforms into another triclinic polymorph: *Wollastonite-II* (Essene 1974; Wo-II), structurally quite similar to *$CaSiO_3$-Walstromite.* Above ~10 GPa the stable form of $CaSiO_3$ is *Ca-Perovskite* (Ca-Pvsk), orthorombic (space group *Pnma*).

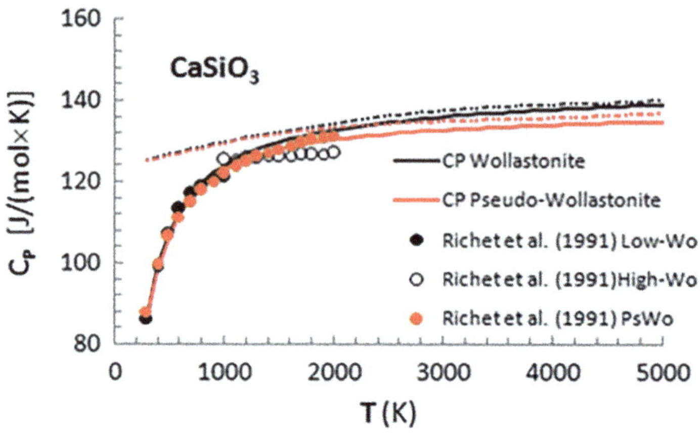


**Fig. 6.29** Isobaric heat capacity of the Wollastonite and Pseudo-Wollastonite. The assessments of Richet et al. (1991) are superimposed for comparative purposes. These are based on the $C_P$ expansions given in their Table 4.1 and limited to the T-range encompassed in their Fig. 4.1. No refitting of the heat capacity function was deemed necessary to conform strictly to the Dulong-Petit asymptotic limit (dotted lines) because the two polymorphs are extrinsically unstable at high-T - high-P conditions

For Wo and PsWo, Belmonte et al. (2017a) adopted the heat capacity coefficients of Berman (1988), which are based essentially on the calorimetric measurements of Krupka et al. (1985a,b), Wagner (1932), White (1919), and Elsner von Gronow and Schwiete (1933): The Low-Wollastonite to High-Wollastonite transition was disregarded. We retain their choice here. We note here, however, in passing, that below ~1000 K the *Low-Wollastonite* data of Richet et al. (1991) are consistent with the selected heat capacity, and a good consistency is observed also for PsWo up to ~2000 K (Fig. 6.29). For $H^{\circ}_{f,298\ S}$ and $S^{\circ}_{f,}$ we initially used the assessment of Berman (1988). In the subsequent optimization, $H^{\circ}_{f}$ of both polymorphs was lowered by 0.51 kJ/mol.

In the following, we will not make any distinction between *Wollastonite-II* (Wo-II) and *$CaSiO_3$-Walstromite.*

The adopted $C_P$ are refinements of the heat capacity equations provided by Swamy and Dubrovinsky (1997), carried out to avoid inconsistencies with the Dulong-Petit limit at high-T (see also Belmonte et al. 2017a) (Table 4.1).

For Wo-II the adopted $V^{\circ}_{298.15}$ is that one of Chatterjee et al. (1994) and $\alpha$ is from Swamy and Dubrovinsky (1994). $K_0$, $K'$, $K_{1,}$ and $K_2$ were refined in this study on the basis of the Wo/Wo-II subsolidus univariant curve (see Fig. 6.30). For Ca-Pvsk $V^{\circ}_{298.15}$ is from Wang et al. (1996); $\alpha$, $K_0$, $K'$, $K_{1,}$ and $K_2$ are those of Belmonte et al. (2017a) (Table 4.2).

The phase diagram of $CaSiO_3$ based on the assessment of Belmonte et al. (2017a) is compared in Fig. 6.30 with the experiments of Gasparik et al. (1994) and Kanzaki et al. (1991) in the P range 8–15 GPa. The agreement is overall satisfactory. To

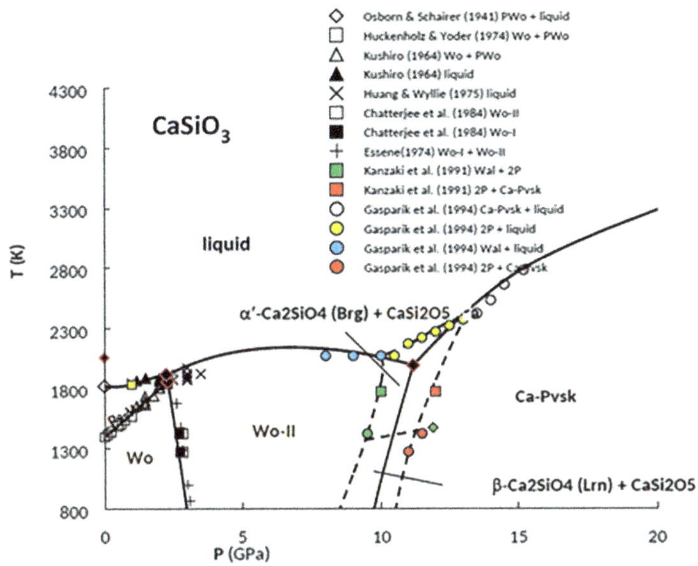


**Fig. 6.30** Computed phase equilibria in the system $CaSiO_3$. Thin solid lines delineate the primary phase fields at subsolidus. The heavy solid line is the melting curve of the substance. Selected experimental data are shown for comparison. The low-P subsolidus ($P \leq 3$ GPa) is consistent with the data of Osborn and Schairer (1941), Kushiro (1964), and Essene (1974). The phase diagram is nearly similar to what was proposed by Gasparik et al. (1994). Here, however, the two phases are identified as β-$Ca_2SiO_4$ and $CaSi_2O_5$ moreover, Walstromite is replaced with Wo-II. From Belmonte et al. (2017a) with modifications

reproduce the liquidus at high pressure, a small excess volume was introduced in the HPA liquid (Table 4.7).

#### 6.2.1.3 $Ca_3SiO_5$

Belmonte et al. (2017a) adopted for *hatrurite* the $H^{\circ}_{f,298.15}$, and $S^{\circ}_{298.15}$ of Barin (1995), with the heat capacity of Berman and Brown (1985). $V^{\circ}_{298.15}$ was from Robie and Hemingway (1995). With respect to a previous optimization (Ottonello et al. 2013), Belmonte et al. (2017a) tuned the thermal expansion α (initially based on Skinner 1966, compilation) to achieve consistency with the Dulong-Petit asymptotic limit (Fig. 6.31).

#### 6.2.1.4 $Ca_3Si_2O_7$

Belmonte et al. (2017a) adopted for *rankinite* ($Ca_3Si_2O_7$) a $H^{\circ}_{f,298.15}$ value nearly similar to what was adopted by Eriksson et al. (1994), while adopting $S^{\circ}_{298.15}$ of Barin (1989) and the $C_P$ function of Berman and Brown (1985). $V^{\circ}_{298.15}$, $K_0$, and K'

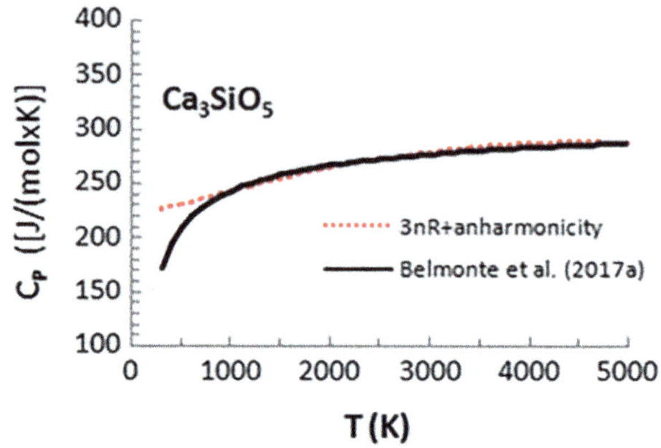


**Fig. 6.31** Isobaric heat capacity of Hatrurite (Berman and Brown 1985). The red-dotted line is shown for comparative purposes to visualize the attainment of the Dulong—Petit asymptotic limit at high-T on the basis of the selected thermophysical parameters (Table 4.2)

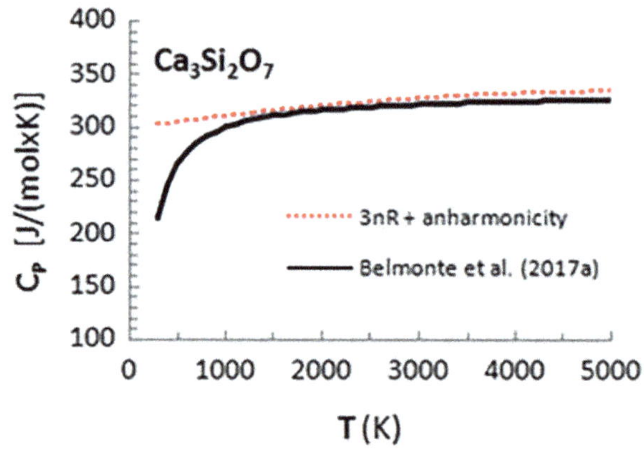


**Fig. 6.32** Isobaric heat capacity of *Rankinite* (Berman and Brown 1985). The red-dotted line is shown for comparative purposes to visualize the level of attainment of the Dulong-Petit asymptotic limit at high-T on the basis of the selected thermophysical parameters (Table 4.2)

were from Holland and Powell (1998). Thermal expansion was initially set according to the estimates of Skinner (1966) and then refined to achieve the correct asymptotic Dulong—Petit limit (not fully respected, however, see Fig. 6.32). We retain here their assessment.

#### 6.2.1.5 $CaSi_2O_5$

*Ca-titanite* ($CaSi_2O_5$) forms by breakdown of Wo-II above ~8 GPa. The adopted $C_P$ is based on the quasi-harmonic lattice dynamics simulation of Swamy and Dubrovinsky (1997) (Table 4.1). The adopted $H°_f$ is the calorimetric value of Schoenitz et al. (2001) (Table 4.1).

$V°_{298.15}$,$K_0$ and K' are taken from Angel (1997) and Angel et al. (1996,1999). The thermal effect on compressibility (i.e., $K_1$ and $K_2$) are those necessary to achieve consistency with the high-T Dulong-Petit asymptotic limit. The remaining parameters are those necessary to reproduce the univariant phase equilibrium boundaries

$CaSi_2O_5$ + β-$Ca_2SiO_4$ = $CaSiO_3$ (Wo-II) and $CaSi_2O_5$ + β-$Ca_2SiO_4$ = $CaSiO_3$ (Ca-pvsk) at subsolidus (Kanzaki et al. 1991; Gasparik et al. 1994) (see Belmonte et al. 2017a, and Fig. 6.30).

### 6.2.2 The $CaO$-$SiO_2$ System at Room Pressure

Various assessments exist in literature (Hillert et al. 1990; Taylor and Dinsdale 1990; Eriksson and Pelton 1993; Wu et al. 1993; Eriksson et al. 1994; Jung et al. 2004,2005). Thermodynamic assessments based on MQC turn out to be consistent with a melting point of lime around 2845 K (Eriksson and Pelton 1993; Wu et al. 1993; Eriksson et al. 1994; Jung et al. 2004,2005). Our assessment, which is based on HPA and the first principles modeling of CaO solid and liquid phases, is consistent with a $T_f$ = 3202 K for lime, in close agreement with recent experimental data (Manara et al. 2014) and MD results (Sun et al. 2010). In the assessment of Belmonte et al. (2017a) (here retained) at room pressure, the $CaO$-$SiO_2$ system has two congruently melting compounds (the orthosilicate $Ca_2SiO_4$ and the metasilicate $CaSiO_3$) and two incongruently melting compounds ($Ca_3SiO_5$ *hatrurite* and $Ca_3Si_2O_7$ *rankinite*) (Fig. 6.33). The main further observations may be resumed as follows:

(1) *α-Larnite* is seen to melt congruently at T = 2404 K (in agreement with experiments; cf. Osborn 1942; Welch and Gutt 1959; Gutt 1968).

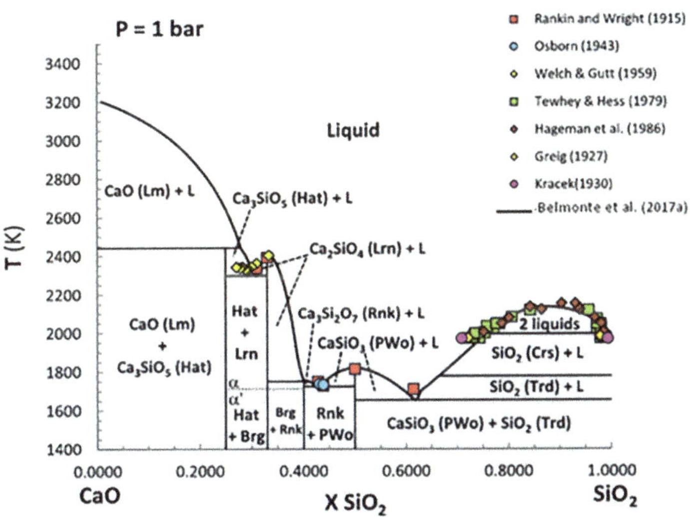


**Fig. 6.33** Melting relations in the $CaO$-$SiO_2$ binary system at P = 1 bar. The HPA liquidus (solid line) is compared with experimental observations. The thin dotted line locates the α/α' phase transition of $Ca_2SiO_4$ at subsolidus. From Belmonte et al. (2017a) with modifications

(2) *Rankinite* ($Ca_3Si_2O_7$) melts incongruently to form larnite ($\alpha$-$Ca_2SiO_4$) plus liquid at T = 1750 K.
(3) A peritectic point (liquid + $\alpha$-$Ca_2SiO_4$ = $Ca_3Si_2O_7$) is observed at $X_{SiO2}$ = 0.41 ± 0.01 T = (1748 ± 5 K, $X_{SiO2}$ = 0.428 according to Rankin and Wright 1915; T = 1737 K, $X_{SiO2}$ = 0.431 according to Osborn, 1943; T = 1745 K and $X_{SiO2}$ = 0.409 according to Hillert et al. 1990; T = 1751 K and $X_{SiO2}$ = 0.409 according to Hillert et al. 1991; T = 1737 ± 10 K and $X_{SiO2}$ = 0.41 ± 0.01 according to Eriksson et al. 1994).
(4) *Rankinite* forms an eutectic with $CaSiO_3$ *PsWo* at T = 1724 K and $X_{SiO2}$ = 0.43, in agreement with experimental findings (Rankin and Wright 1915; Osborn 1943). The calculated melting point of $CaSiO_3$ *PsWo* at 1-bar pressure is 1820 K, which is consistent with the value given by Osborn and Schairer (1941) (i.e., 1821 K). Some discrepancy exists on the location of the eutectic point between $CaSiO_3$ *PsWo* and $SiO_2$ (*High-Tridymite*), which is predicted to occur at T = 1653 K and $X_{SiO2}$ = 0.62 by Belmonte et al. (2017a), while experimental works locate this point at T = 1709 K and $X_{SiO2}$ = 0.614–0.624 (Rankin and Wright 1915; Osborn 1943).
(5) A peritectic point between *Lime* and *Hatrurite* (i.e., liquid + CaO = $Ca_3SiO_5$) is observed at T = 2443 K, $X_{SiO2}$ = 0.28 ± 0.01.
(6) An eutectic point between *Hatrurite* and *α-Larnite* is observed at T = 2298 K, $X_{SiO2}$ = 0.31 ± 0.01.
(7) A liquid–liquid miscibility gap extends from the nearly pure silica composition to $X_{SiO2}$ = 0.76 at P = 1 bar. The monotectic occurs at T = 1994 K (T = 2005 K and $X_{SiO2}$ = 0.752 according to Hageman et al. 1986; T = 1971 K and $X_{SiO2}$ = 0.71–0.72 according to Greig 1927; Kracek 1930; Ol'shanskii 1951; Tewhey and Hess 1979; see also Thompson et al. 2007, for a review of experimental results). The consolute point is predicted to occur at T = 2137 K and $X_{SiO2}$ = 0.87 ± 0.01 (T = 2144 K and $X_{SiO2}$ = 0.91 ± 0.01 according to Tewhey and Hess 1979; T = 2159 ± 1 K and $X_{SiO2}$ = 0.902 according to Hageman et al. 1986)

### 6.2.3 The CaO-$SiO_2$ System at High Pressure

The calculated CaO-$SiO_2$ phase diagram at P = 1 GPa is shown in Fig. 6.34.

The main features may be resumed as follows:

(1) $Ca_3SiO_5$ (*hatrurite*) and $Ca_3Si_2O_7$ (*rankinite*) are no longer stable at the liquidus.
(2) *α-Larnite* melts congruently at T = 2589 K.
(3) There are only three eutectic points: one between CaO (*Lime*) and $Ca_2SiO_4$ (*α-Larnite*) at T = 2418 K and $X_{SiO2}$ = 0.30 (see also Hudon et al. 2005); a second one between $Ca_2SiO_4$ (α'-*Bredigite*) and $CaSiO_3$ (*PsWo*) at T = 1733 K; a third one between *PsWo* and *β-Qtz* at a slightly lower T and higher $SiO_2$ content.

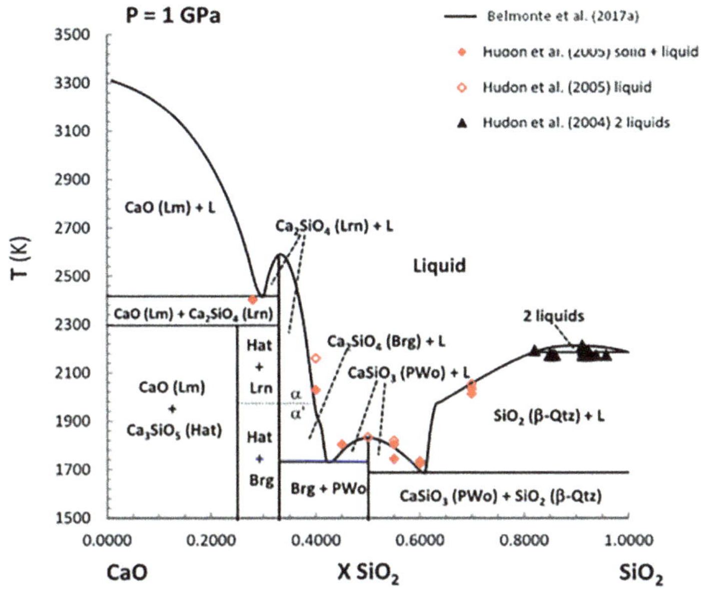


**Fig. 6.34** Melting relations in the $CaO-SiO_2$ binary system at P = 1 GPa, Experimental observations (symbols) (Hudon et al. 2004, are superimposed for comparative purposes. The thin dotted line locates the α/α' phase transition of $Ca_2SiO_4$ at subsolidus. From Belmonte et al. (2017a) with modifications

(4) The liquid–liquid miscibility gap still survives at P = 1 GPa, though flattened (a monotectic-T only 28 K lower than the consolute-T) and drastically reduced in width (see also Hudon et al. (2004, 2005).

The evolution of the $CaO-SiO_2$ liquidus with pressure (from 1 bar to 25 GPa) is shown schematically in Fig. 6.35:

(1) $Ca_2SiO_4$ melts congruently at all investigated pressures, despite the structural rearrangements dictated by pressure (see Fig. 6.28).
(2) $CaSiO_3$ melts congruently as *PsWo* in the low-pressure range, then melts incongruently as *Wollastonite-II* (or *Walstromite*) in the intermediate-pressure range, and finally again congruently as *Ca-Perovskite* at higher pressures.
(3) $Ca_2SiO_4$ (an efficient thermal barrier in the whole investigated range) forms two invariant points (eutectic with CaO on the left-hand side of the phase diagram and peritectic with $CaSiO_3$ on the right-hand side).
(4) In a narrow pressure range (between ~10 and 15 GPa), the metasilicate is no more stable on the liquidus, and the two-phase assemblage $Ca_2SiO_4$ (*α'-Bredigite* first, then *β-Larnite* + $CaSi_2O_5$, instead of $CaSiO_3$) becomes stable.
(5) The composition of the eutectic point between CaO (*Lime*), $Ca_2SiO_4$ (*α-Larnite* or *α'-Bredigite*), and liquid shifts from $X_{SiO2} = 0.30$ at P = 1 GPa (and T = 2418 K) to $X_{SiO2} = 0.22$ at P = 10 GPa (and T = 3321 K), then its silica content slowly increases again up to $X_{SiO2} = 0.24$ at P = 25 GPa (and T = 3945 K).

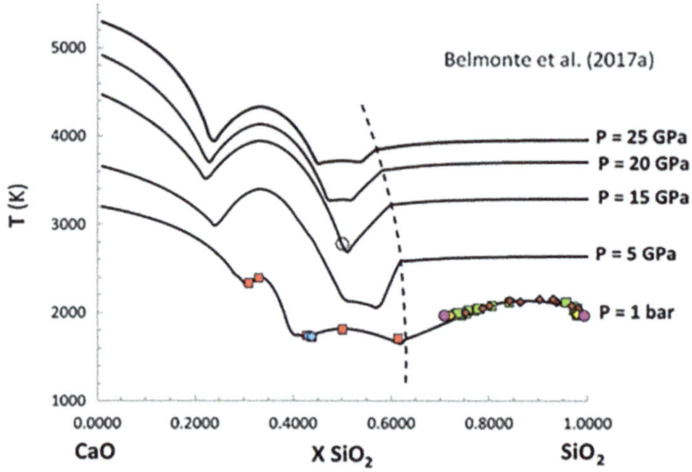


**Fig. 6.35** Evolution of the CaO-$SiO_2$ liquidus with pressure (from 1 bar to 25 GPa). Experimental observations (symbols) are superimposed for comparative purposes. Symbols as in Fig. 6.33, except for Gasparik et al. (1994) at P = 15 GPa (empty circle). The dashed line indicates the limit of maximum depolymerization in the liquid (HPA model) at different pressures. From Belmonte et al. (2017a) with modifications

(6) The same trend occurs in the eutectic between $CaSiO_3$ and $SiO_2$: the silica content of this eutectic lowers with the increase of P, from $X_{SiO2} = 0.62$ at P = 1 bar to $X_{SiO2} = 0.49$ at P = 18 GPa, then increases again up to $X_{SiO2} = 0.53$ in the pressure range 18 GPa < P < 25 GPa.

(7) The liquid–liquid miscibility gap disappears between 1 and 2 GPa (Figs. 6.34 and 6.35). The attainment of the limit of maximum depolymerization in the liquid produces a kink in the liquidus curve on the $SiO_2$-rich side of the phase diagram (dashed line in Fig. 6.35), which is possibly a numerical artifact of the computational procedure.

## 6.3 MgO-$Al_2O_3$

Although $Al_2O_3$ is less "*acidic*" than $SiO_2$, the NM-NF polymeric interaction parameters (A and B in Table 4.7) are quite similar in the MgO-$SiO_2$ and MgO-$Al_2O_3$ joins when replacing $SiO_2$ with $Al_2O_3$. Namely, A changes from −3330 to −3350, and the B term is enhanced from −2.5 K to −1.74 K. We have again an indirect confirmation of the fundamental correctness of the polymeric model, comparing the enthalpy of mixing with the value obtained from the enthalpy of fusion of the stoichiometric spinel compound $MgAl_2O_4$. Barin (1995) gives in fact $\Delta H_{fusion} = 196.648$ kJ/mol at $T_f = 2408$ K, which is not far from the NIST-JANAF value (i.e., $\Delta H_{fusion} = 194.6 \pm 20.9$ kJ/mol; Chase, 1998). These values, based on Eq. 6.1, return an enthalpy

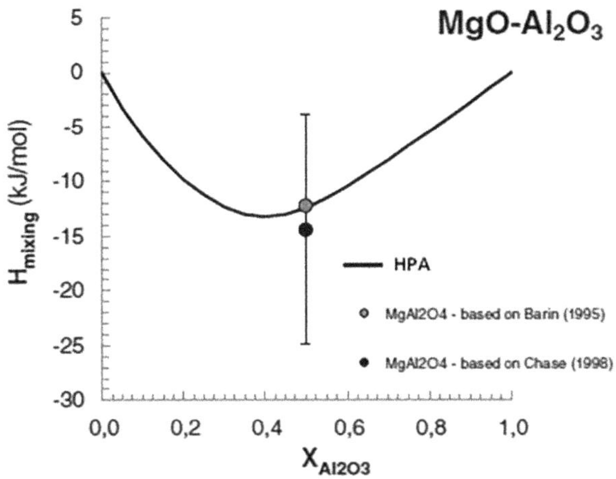


**Fig. 6.36** Enthalpy of mixing in the $MgO-Al_2O_3$ system. Estimates based on the experimental enthalpies of fusion of stoichiometric spinel $MgA_2O_4$ (Barin 1995; Chase 1998) are compared with HPA model calculations (curve in bold). Modified from Belmonte et al. (2017b)

of mixing of $-12.3$ kJ/mol and $-14.4 \pm 9.6$ kJ/mol, respectively, for the $MgAl_2O_4$ liquid composition, in overall agreement with the model predictions (Fig. 6.36).

The two limiting pure components MgO and $Al_2O_3$, do not undergo any phase transition at $P = 1$ bar, and the only intermediate compound that forms in the system is the spinel solid mixture. For stoichiometric spinel, we adopt the enthalpy of formation from the elements, entropy, isobaric heat capacity, and standard-state volume after Berman (1988). For the thermoelastic parameters, Belmonte et al. (2017b) adopted the parameterization of Saxena et al. (1993). Subtracting the anharmonicity contribution based on these parameters from $C_P$, the obtained isochoric heat capacity is, however, inconsistent with the Dulond-Petit limit. We introduce here slight modifications to thermal expansion and $K = f(T)$ in order to obey, at best, first principles (Fig. 6.37 and Table 4.2).

According to Alper et al. (1962), spinel ($MgAl_2O_4$) melts congruently at 2381 K and forms an eutectic with MgO at $T = 2271$ K and $X_{Al2O3} = 0.325$. $MgAl_2O_4$ forms, however, at subsolidus, a solid mixture with the $Al_2O_3$ component. The solubility is almost complete. According to Viechnicki et al. (1974), an $Al_2O_3$-rich $MgAl_2O_4$-$Al_2O_3$ solid mixture ($X_{Al2O3} = 0.85$) forms an eutectic with corundum at 2248 K and $X_{Al2O3} = 0.90$. According to Roy et al. (1953), Saafeld and Jagodzinski (1957), Lejus and Collongues (1962), Lejus (1964), Shirasuka and Yamaguchi (1974), at this temperature, the solubility of alumina in the solid mixture is higher, and the existence of an eutectic is debatable. The spinel solid mixture has been modeled by Mao et al. (2005a) in the framework of the two-sublattice compound energy formalism (CEF) of Hillert et al. (1988). A four-sublattice CEF for the same substance has been more recently proposed by Zienert and Fabrichnaya (2013). Both models are obviously quite successful in depicting both the internal disorder as a function of T and the melting behavior due to the adopted computational freedom. Belmonte et al. (2017b) preferred to treat the energy properties of the spinel solid mixture as a simple two-component phase with mixing on one site. Here we adopted the same conceptual model but refined the mixing parameters as detailed at Par. 4.4.2.

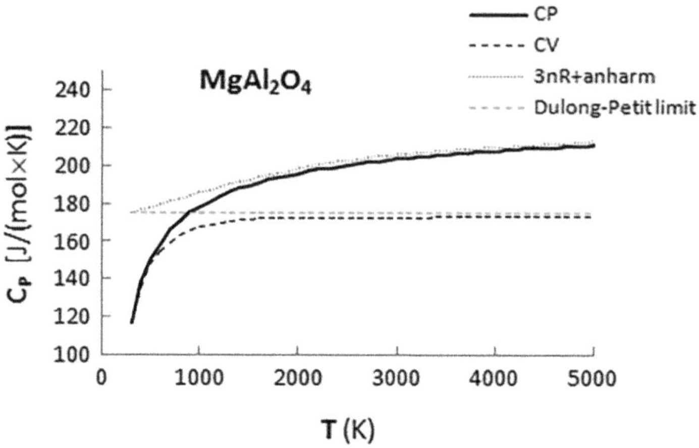


**Fig. 6.37** Isobaric (heavy black solid line) and isochoric (black dashed line) of stoichiometric spinel. $C_P$ after Berman (1988), $C_V$ obtained in this study from inverse modeling of anharmonicity

Accounting for the solubility of alumina in spinel and magnesium oxide has some effect on the topology of the $MgO$-$Al_2O_3$ liquidus curve at 1-bar pressure (Fig. 6.38). If solid-state mixing effects in solid phases are neglected, the experimental results of Roy et al. (1953) are well reproduced, and a eutectic between pure spinel and corundum is observed at T = 2296 K and $X_{Al2O3} = 0.83$. The eutectic between periclase and spinel is predicted to occur at T = 2203 K and $X_{Al2O3} = 0.31$ if only pure compounds are considered, while it is displaced to T = 2258 K and $X_{Al2O3} = 0.325$ when solubility of alumina in both magnesium oxide and spinel is taken into account. Finally, while pure $MgAl_2O_4$ turns out to melt at T = 2383 K, the spinel solid mixture shows a flat maximum at T = 2398 K between $X_{Al2O3} = 0.465$ and $X_{Al2O3} = 0.585$.

Increasing the pressure from 1 bar to 25 GPa leads to a sub-parallel upward shift of the liquidus curve and to a progressive shrinking of the primary phase field of spinel, which is no longer present at P higher than ~12 GPa (Fig. 6.38).

## 6.4 CaO-$Al_2O_3$

This joint is of high importance in ceramurgy because many important phases, present in the Portland Cement, nucleate in it. The presence of $SiO_2$ and MgO limits at all P the primary phase field of these substances, so that the importance of this join in CMAS is ancillary at best (see later). Notwithstanding this fact, it must be stressed that the majority of solids nucleating in the CaO-$Al_2O_3$ system are commonly observed in meteorites, and one of them (hibonite) is an important witness of the first moments of solar condensation events.

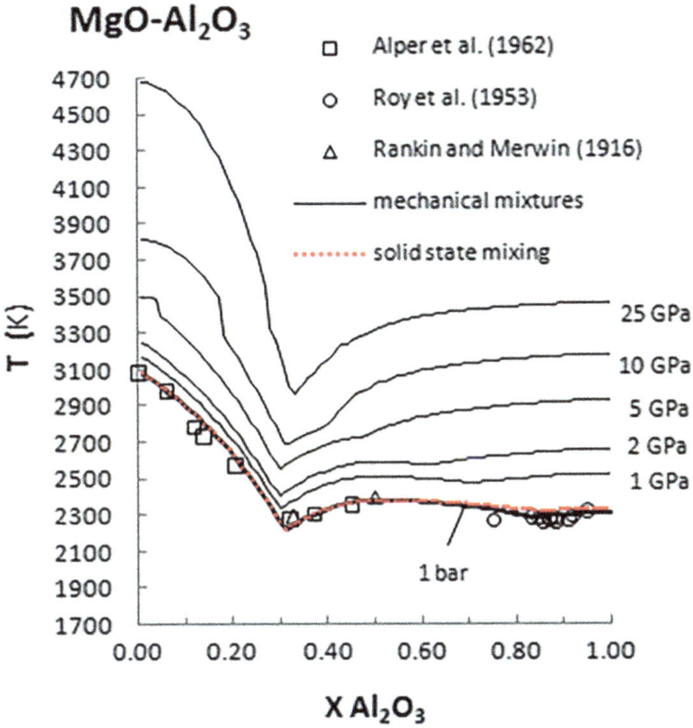


**Fig. 6.38** HPA predictions at room pressure are compared with the experimental observations of Alper et al. (1962), Roy et al. (1953), and Rankin and Merwin (1916). The increase in pressure rises progressively the melting temperature as expected, but the primary phase field of spinel shrinks progressively and disappears at some pressure above 5 GPa. The effect of solid-state mixing (spinel solid mixture and oxide solid mixture) is almost negligible (red-dashed line)

In a previous investigation (Ottonello et al. 2013), we assigned to the polymeric interaction parameter A a value of −9809.5, with B set to zero. In this study, we raise A to −10,000, posing B = −1.6. The strain energy (previously conformed through 12 parameters) is sufficiently well accounted for by four terms (Table 4.7). As already noted by Eriksson and Pelton (1993), in order to reproduce the very low eutectic temperature of the system in the framework of the MQC model, a "*large negative excess entropy*" must be introduced. In the HPA model, this entropic effect is accounted for through four large entropic strain contributions extended over the entire NM-NF system (Table 4.7). Indeed, as observed by Drewitt et al. (2017) the atomic-scale structure of aerodynamically levitated and laser-heated liquid tricalcium aluminate ($Ca_3Al_2O_6$) [measured at 2073 K by using the method of neutron diffraction with Ca isotope substitution (NDIS)] indicates, through Molecular Dynamics (MD) simulation and reverse Monte Carlo (RMC) refinement methods, the dominant presence of $AlO_4$ tetrahedra in an infinite network, with a slightly higher concentration of bridging oxygens than expected, around 10% of $AlO_4$ monomers, and $Al_2O_7$

dimers. At this composition, in the absence of strain contributions, the HPA model would predict complete depolimerization. In this peculiar case, the strain energy thus apparently has a "*restoring effect*" fighting depolymerization.

### 6.4.1 Intermediate Crystalline Compounds

The compound $Ca_{12}Al_{14}O_{33}$, previously identified by Lagerqvist et al. (1937) by an X-ray investigation, was later shown by Nurse et al. (1965), using high-temperature microscopy under controlled atmosphere, to be a hydrate of formula $Ca_{12}Al_{14}0_{32}(OH)_2$, and, as such, not present in the nominally anhydrous binary CaO-$A1_2O_3$ join. Besides the two limiting pure components whose properties have been discussed in the previous paragraphs, the other four crystalline phases form in the system: $Ca_3Al_2O_6$, $CaAl_2O_4$, $CaAl_4O_7$ (grossite), and $CaAl_{12}O_{19}$ (hibonite).

#### 6.4.1.1 $Ca_3Al_2O_6$

Tricalcium aluminate is an important phase in the Portland Cement (Kosmatka 2012). It belongs to the Mayenite Supergroup and crystallizes in the cubic system with space group *Pnmn* or, more probably, *Pa3* (Mondal and Jeffery 1975). The unit cell contains eight cyclic $Al_6O_{18}{}^{18-}$ anions, actually formed by six interconnected (corner-sharing) $AlO_4$ tetrahedra. The thermodynamic properties of the substance adopted in this study are listed in Table 4.1. $H^0{}_{f,298.15}$ is from Coughlin (1956), $S^0{}_{298.15}$ is from King (1955), and the heat capacity function is from Berman and Brown (1985). Concerning the thermophysical properties (Table 4.2), the standard-state molar volume $V^0{}_{298.15}$ is from Mondal and Jefffery (1975), the thermal expansion function is after Skinner (1966) (refitted by Ottonello et al. 2013). Since we have no hints on the thermal effect on the bulk modulus, we obtained it by inverse methods, stemming from anharmonicity (constraining obedience to the Dulong-Petit limit at high-T). To do this, we adopted the *ab-initio* bulk modulus $K_0 = 102.9$ GPa of Manzano et al. (2009), posing $K_1 = -292.2$ bar/K, and $K_2 = 0.0225$ bar $\times$ $K^{-2}$. The results of this exercise are visualized in Fig. 6.39

#### 6.4.1.2 $CaAl_2O_4$

Calcium aluminate occurs in two forms: a low-P monoclinic phase (space group *P21/n*), known as krotite, found in meteorites, and dmitryivanovite, also monoclinic (space group *P*21/b), and also found in meteorites. In the following, we will make no distinction between the two polymorphs.

The thermodynamic properties of the substance adopted in this study are listed in Table 4.1. $H^0{}_{f,298.15}$ is from Coughlin (1956), $S^0{}_{298.15}$ is from King (1955), and the heat capacity function is from Berman and Brown (1985). Concerning the thermophysical

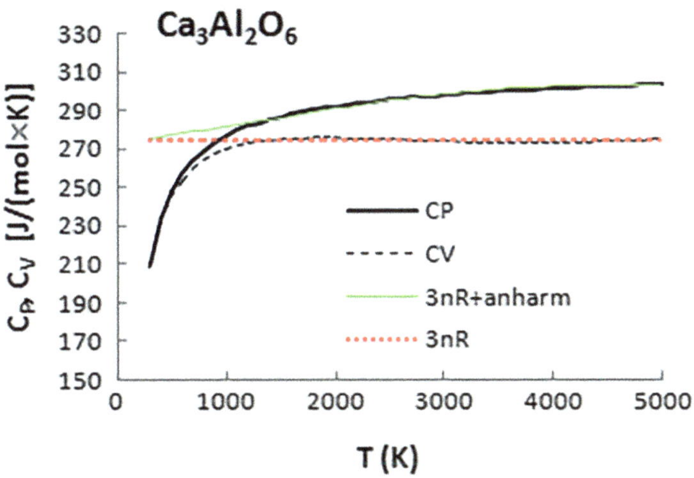


**Fig. 6.39** Isobaric (heavy black solid line) and isochoric (black dashed line) of $Ca_3Al_2O_6$. $C_P$ after Berman and Brown (1985), $C_V$ obtained in this study from inverse modeling of anharmonicity

properties (Table 4.2), the standard-state molar volume $V^0_{298.15,}$ the bulk modulus $K_{0,}$ and the baric derivative K' were taken from Akaogi et al. (1999), Kojitani et al. (2003), and Ono et al. (2009) (Ottonello et al. 2013). Because the anharmonicity of the substance is quite marked, we adopted a constant thermal expansion slightly higher than the function previously obtained by us (Ottonello et al. 2013) on the basis of the tabulation of Skinner (1966) ($\alpha 1 = 2.8 \times 10^{-5}$ $K^{-1}$). The thermal effect of the bulk modulus was refined in this study in order to achieve compliance with the Dulong-Petit limit ($K_1 = 102.3$ bar/K, $K_2 = -0.048$ bar/$K^2$) (Fig. 6.40).

#### 6.4.1.3 $CaAl_4O_7$

Calcium dialuminate (also known as grossite) is monoclinic (space group *B2/b*). It occurs in tectonic collisional zones (Dellagiustaite-type localities; Camara et al. 2019), and it is observed also in meteorites. The thermodynamic properties of the substance adopted in this study are listed in Table 4.1. $H^0_{f,298.15}$ is from Kubaschevski et al. (1967), $S^0_{298.15}$ is from King (1955), and the heat capacity function is from Berman and Brown (1985). Concerning the thermophysical properties (Table 4. 2), the standard-state molar volume $V^0_{298.15}$ is from Goodwin and Lindtop (1970). Hofmeister (2004), through vibrational spectroscopy, obtained a bulk modulus $K_0$ = 200 GPa and a baric derivative K' = 4. A Molecular Dynamics (MD) simulation gave a constant thermal expansion $\alpha_1 = 2.754 \times 10^{-5}$ $K^{-1}$ (Suzuki and Ohji 2004). We assume in this study a thermal effect $K_1 = -466.7$ bar/K, $K_2 = 0.036$ bar/$K^2$ in order to obtain a reasonable isocoric heat capacity (Fig. 6.41).

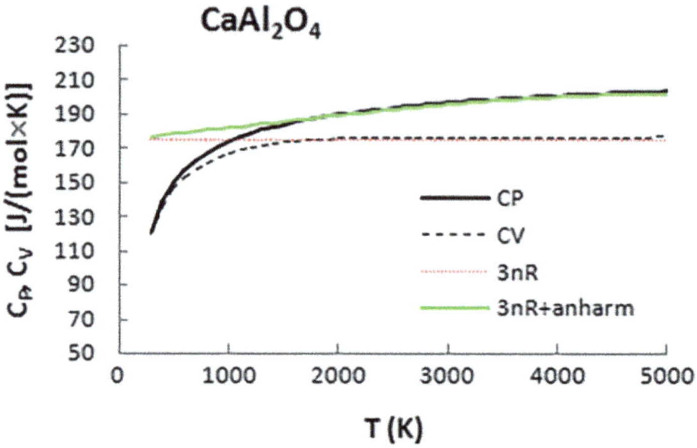


**Fig. 6.40** Isobaric (heavy black solid line) and isochoric (black dashed line) heat capacities of $CaAl_2O_4$. $C_P$ after Berman and Brown (1985), $C_V$ obtained in this study from inverse modeling of anharmonicity

**Fig. 6.41** Isobaric (heavy black solid line) and isochoric (black dashed line). Heat capacities of $CaAl_4O_7$: $C_P$ after Berman and Brown (1985), $C_V$ obtained in this study from inverse modeling of anharmonicity

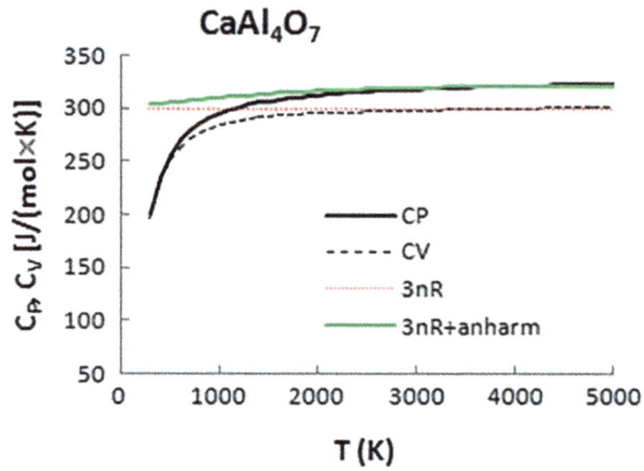


### 6.4.1.4 $CaAl_{12}O_{19}$

$CaAl_{12}O_{19}$ (hibonite) belongs to the plumboferrite group (spatial group *P* $6_3/mmc$). Hibonite crystals older than 4.5 Ga have been observed in the Murchison meteorite. We adopt for the thermodynamic properties of this substance the values suggested by Eriksson and Pelton (1993) (Table 4.1). For the standard-state molar volume, we adopt the value indicated by Kato and Saalfeld (1968) ($V^0_{298.15} = 176.729$). For $K_0$ Hofmeister (2004), by fixing K' to 4, obtained $K_0 = 230$ GPa. The same Author indicated a constant thermal expansion $\alpha_1 = 2.1 \times 10^{-5}\ K^{-1}$. We raised in this study the thermal expansion to $\alpha_1 = 2.8 \times 10^{-5}\ K^{-1}$ and adopted a thermal effect $K_1 =$

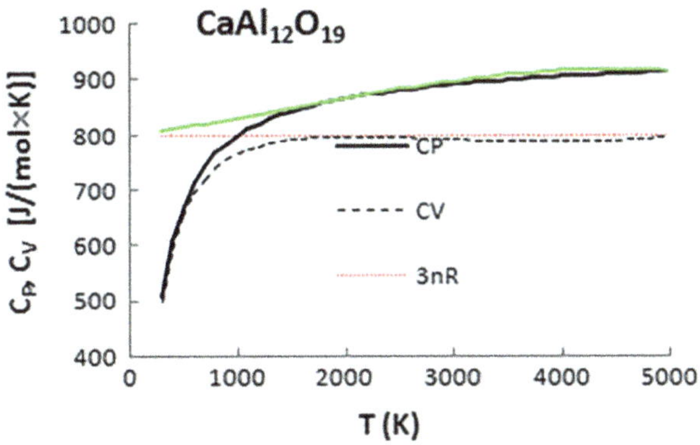


**Fig. 6.42** Isobaric (heavy black solid line) and isochoric (black dashed line) heat capacities of $CaAl_{12}O_{19}$. $C_P$ after Eriksson and Pelton (1993). $C_V$ obtained in this study from inverse modeling of anharmonicity

190 bar/K, $K_2 = -0.07$ bar/$K^2$ in order to obtain a reasonable isocoric heat capacity (Fig. 6.42).

## 6.4.2 *The CaO-$Al_2O_3$ Binary*

Nurse et al. (1965) reported all four compounds to melt incongruently: $Ca_3Al_2O_6$ at 1814 K and $X_{Al2O3} = 0.74$; $CaAl_2O_4$ at 1877 K and $X_{Al2O3} = 0.49$; $CaAl_4O_7$ at 2038 K and $X_{Al2O3} = 0.67$; $CaAl_{12}O_{19}$ at 2106 K and $X_{Al2O3} = 0.74$. He observes here a deep eutectic at 1635 K and $X_{Al2O3} = 0.36$. Eutectic temperature and composition were later confirmed by Nityanand and Fine (1983). According to Rolin and Pham (1965), $CaAl_2O_4$ melts congruently at 1875 K. According to Wisnyi (1955), $CaAl_4O_7$ melts congruently at 2021 K. Although somewhat contradictory, all the above observations delineate a reasonable liquidus (Fig. 6.43).

The eutectic of Nurse et al. (1965) is reproduced by HPA at T = 1601 K and $X_{Al2O3} = 0.34$. Congruent melting of $CaAl_2O_4$ is predicted to occur at T = 1880 K. $CaAl_4O_7$ is predicted to melt congruently at 1995. According to HPA, with the increase of pressure, the melting of $CaAl_2O_4$ remains congruent, the eutectic between $CaAl_2O_4$ and $Ca_3Al_2O_6$ is progressively shifted at lower $X_{Al2O3}$, and an eutectic between $CaAl_2O_4$ and $CaAl_4O_7$ appears in the acidic part of the join (Fig. 6.43).

## 6.5 $Al_2O_3$-$SiO_2$

This join is of high importance in metamorphic petrology due to the presence in the system of three alumino-silicate polymorphs ($Al_2SiO_5$ in the forms andalusite, kyanite, sillimanite), indicators for the pressure and temperature formation conditions in metamorphic sediments. Besides the three aluminosilicates, one observes the formation of mullite ($Al_6Si_2O_{13}$). Notwithstanding this topological complexity, the

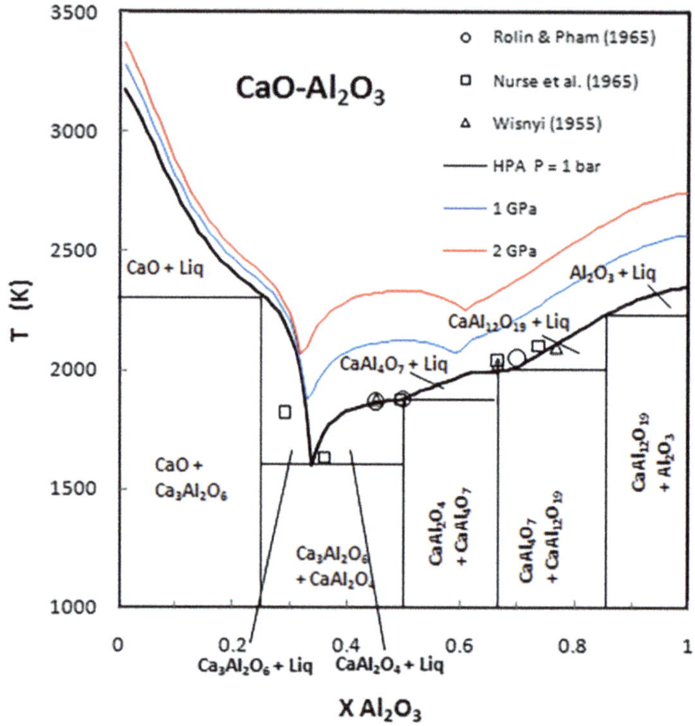


**Fig. 6.43** Model predictions for the CaO-$Al_2O_3$ join at P = 1 bar, compared with experimental observations. HPA Liquidi at 1 and 2 Gpa are also delineated

HPA is able to represent the Gibbs free energy of mixing through a single entropic parameter (B = −1.85; Table 4.7) and two entropic strain contributions, necessary to conform the eutectic at high $SiO_2$ content.

## *6.5.1 Intermediate Crystalline Compounds*

### 6.5.1.1 $Al_6Si_2O_{13}$

Stoichiometric mullite ($Al_6Si_2O_{13}$) is an orthorombic compound (space group *Pbam*) first identified by Bowen and Greig (1924) and considered to melt incongruently (Skola 1937; Shears and Archibald 1954; Konopicky 1956; Gelsdorf et al. 1958). Other studies described a congruent melting behavior (e.g., Aramaki and Roy 1962; Staronka et al. 1968), while for some Authors the melting behavior depends in a complex manner upon the cooling rate and non-stoichiometry (e.g., Chaudhuri 1987). More recently, Mao et al. (2005b) reproduced the 1-bar subsolidus phase relations in the $Al_2O_3$-$SiO_2$ system by adopting a four-sublattice solution model, as anticipated by Hillert et al. (1989), with a Redlich–Kister parameterization of interactions at subsolidus. We preferred to model the energy properties of the mullite phase as a nonideal subregular mixture of the fictive isostructural (i.e., orthorombic) $Al_2O_3$ and $SiO_2$ end-members, as already proposed long ago by Howald and Eliezer (1978). Details on the mixing model adopted in this work for mullite have been given at Par. 4.4.1. According to our simple model, the melting behavior of the mullite solid mixture at P = 1 bar is in good agreement with the observations of Klug et al. (1987) (see later).

#### 6.5.1.2 $Al_2SiO_5$ Polymorphs

The three polymorphs differ substantially from each other based on the coordination state of aluminum in the unit cell: half of it is always octahedrally coordinated, but half is six-, five-, and four-coordinated in kyanite, andalusite, and sillimanite, respectively.

##### 6.5.1.2.1 Andalusite

This $Al_2SiO_5$ low-T–low-P polymorph is orthorhombic (space group *Pnnm*). The crystal structure of andalusite consists of tetragonally distorted Al- octahedra sharing edges to form chains along the c-axis. These octahedral chains are linked through $SiO_4$ tetrahedra and edge-sharing pairs of Al-trigonal bipyramids (Ralph et al. 1984, and references therein). We adopt in this study for this phase (and the other two polymorphs) the thermodynamic parameters of Berman (1988). Concerning the thermoelastic properties of the substance, Berman (1988) adopted a simple linear parameterization of the type:

$$\frac{V_{P,T}}{V^0_{P_r,T_r}} = 1 + \upsilon_1(P - P_r) + \upsilon_2(P - P_r)^2 + \upsilon_3(T - T_r) + \upsilon_4(T - T_r)^2 \quad (6.4)$$

With little effort, adopting the $\upsilon_1, \upsilon_2, \upsilon_3, \upsilon_4$ coefficients indicated by Berman (1988), one can easily retrieve $\alpha = \frac{1}{V}\left(\frac{\partial V}{\partial T}\right)$ and $K = \left[-\frac{1}{V}\left(\frac{\partial V}{\partial P}\right)\right]^{-1}$ coefficients valid within the indicated validity range. The bulk modulus so obtained is 133.5 GPa (almost constant over T) and the thermal expansion is simply linear with $\alpha = \alpha_o T + \alpha_1 = 0.7161 \times 10^{-9} \times T + 2.336 \times 10^{-5}$.. The obtained bulk modulus ($K_0 = 133.5$ GPa) is consistent with the experimental results of Ralph et al. (1984) (i.e., $K_0 = 135 \pm 0.10$ GPa, K' = 4. The thermal expansion is in line with the power diffraction experiments of Hu et al. (2018), which found a constant α up to 1000 °C ($\alpha_1 = 2.55 \times 10^{-5}\ K^{-1}$). A modern investigation carried out by Fortes (2019) using time-of-flight neutron powder diffraction from 10 to 1573 K revealed a complex thermal behavior with formerly unknown regions of substantial negative linear expansivity below room temperature. Notwithstanding this complexity, at high-T, the thermal expansion is

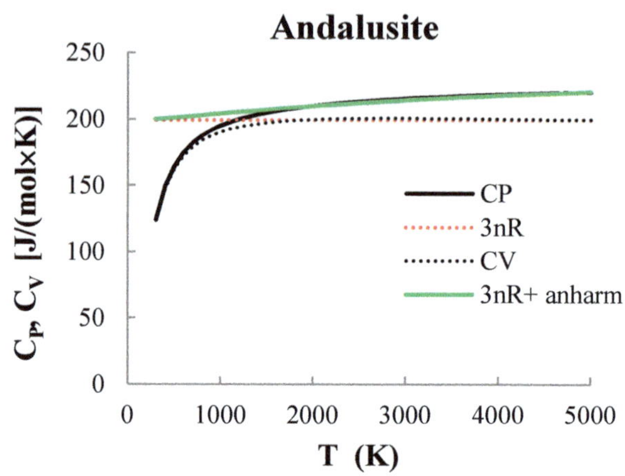


**Fig. 6.44** Isobaric (heavy black solid line) and isochoric (black dashed line) heat capacities of andalusite. $C_P$ after Berman (1988), $C_V$ obtained in this study

consistent with Hu et al. (2018) and Berman (1988) and is here preferred because it extends to higher T.

Adopting $K_0 = 139$ GPa K' $= 4$ and the thermal expansion retrieved from Fortes (2019) and arbitrarily extended to higher T (Table 4.2), it is sufficient to admit a quite limited thermal dependence of the compressibility (i.e., $K_1 = -78$ bar/K) to obtain a reasonable isocoric heat capacity in a wide T-range (Fig. 6.44).

#### 6.5.1.2.2 Kyanite

*Kyanite* is triclinic (space group $P\overline{1}$). It can be considered as a distorted cubic close-packed arrangement of oxygen O atoms, with 10% of the tetrahedral sites filled with Si and 40% of the octahedral sites filled with Al (Yang et al. 1997a). The $AlO_6$ octahedra form zigzag edge-sharing chains structures cross-linked by alternating with $SiO_4$ tetrahedra (Yang et al. 1997a). Proceeding as done for *andalusite,* we obtain, stemming from the linear parameterization of Berman (1988), a thermal expansion $\alpha = \alpha_o T + \alpha_1 = 0.9468 \times 10^{-9} \times T + 2.709 \times 10^{-5}$ (here adopted). At low-T this thermal expansion is consistent with the parameters that may be regressed from the volumetric data of Skinner (1966) (i.e. $\alpha_0 = 1.015 \times 10^{-8}$; $\alpha_1 = 1.964 \times 10^{-5}$; T $\leq 1000$ °C) but the increase of α with increasing T is much more limited and quite consistent with the observations of Fortes (2019) over a wide T range.

The bulk modulus obtained with Berman's linear parameters is K = 154.3, while the experimental results of Yang et al. (1997a) give a lower compressibility (i.e., $K_0 = 193.0$ GPa, K' $= 4$), confirmed by the later observations of Friedrich et al. (2004) (i.e., $K_0 = 190 \pm 11$ GPa, K' $= 4$). To obtain an isocoric heat capacity approximately consistent with the Dulong-Pertit limit and adopting the experimental bulk modulus of Yang et al. (1997) at the athermal limit and the thermal expansion derived from the model of Berman (1988) (Table 4.2) we arrive to the conclusion that the isobaric heat capacity of Berman (1988) is slightly overestimated above 1500 K (i.e. above

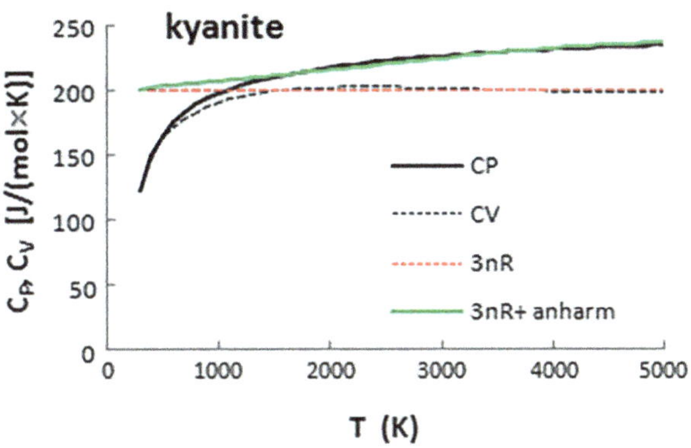


**Fig. 6.45** Isobaric (heavy black solid line) and isochoric (black dashed line) heat capacities of kyanite. $C_P$ after Berman (1988), $C_V$ obtained in this study from inverse modeling of anharmonicity., based on the thermal expansion of Berman (1988) and the compressibility of Yang et al. (1997a)

his observational region, the calorimetric data of Pankratz and Kelley 1964). To reduce this bias, we must assume complex thermal effects on compressibility, i.e., a second-order thermal dependence of K with $K_1 = 153$ bar/K, $K_2 = -0.051$ bar/$K^2$ (Fig. 6.45).

#### 6.5.1.2.3 Sillimanite

*Sillimanite* is orthorombic (space group *Pnma*). The bulk modulus obtained with Berman's linear parameters is K = 132.3, while the experimental results of Yang et al. (1997b) give again a somewhat lower compressibility (i.e., $K_0 = 171.0$ GPa, K' = 4), and those of Burt et al. (2006) are $K_0 = 164$ GPa with K' at 5.0. The thermal expansion experiments of Hu et al. (2018) valid below T ≈ 1300 K delineate a linear increase of α with T, with $\alpha_0 = 7.1 \times 10^{-9}$ and $\alpha_1 = 1.407 \times 10^{-5}$K. From the data tabulated by Skinner (1966) we retrieve a more consistent T effect (i.e. $\alpha_0 = 4.192 \times 10^{-6}$ and $\alpha_1 = 2.015 \times 10^{-8}$ K), while from the linear approach of Berman (1988) we obtain a negative effect (i.e. $\alpha_0 = -1.696 \times 10^{-10}$ and $\alpha_1 = 1.347 \times 10^{-5}$ K). The thermal expansion of Fortes (2019) is again intermediate between Berman's and Hu et al. (2018) and is here preferred (see the third-order expansion in Table 4.2). The isocoric heat capacity is approximately consistent with the Dulong-Petit limit when adopting the data of Yang et al. (1997b) and Hu et al. (2018). The thermal effect on T ($K_1 = -257$ bar/K) was obtained through inverse modeling of anharmonicity (Fig. 6.46).

As we may see in Fig. 6.47, the developed thermodynamic-thermophysical parameterization results in subsolidus equilibria consistent with the reversal experiments. The triple point is seen to occur at P = 3.75 kbar and T = 510 °C. This is quite consistent with the reversal of the kyanite-sillimanite inversion determined hydrothermally by Newton (1966a) and with the calculated slope at this point (17.7 ± 1.0 bars per Celsius).

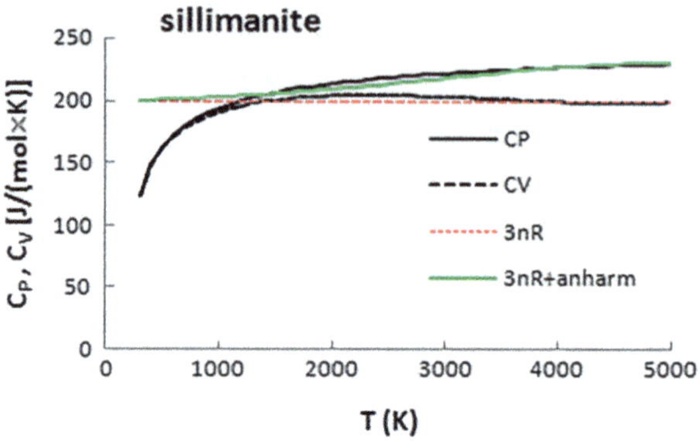


**Fig. 6.46** Isobaric (heavy black solid line) and isochoric (black dashed line) heat capacities of sillimanite. $C_P$ after Berman (1988), $C_V$ obtained in this study from inverse modeling of anharmonicity, based on the thermophysical parameters of Hu et al. (2018) and (Yang et al. 1997b)

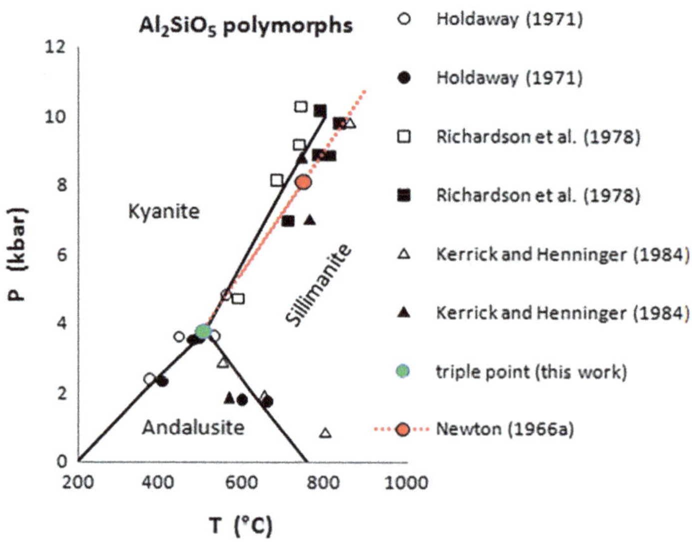


**Fig. 6.47** Polymorphic transitions at subsolidus for $Al_2SiO_5$. The solid lines are the extrinsic stability limits obtained in this study. Open and filled symbols are reversal experiments after Holdaway (1971) (circles), Kerrick and Henninger (1984) (triangles), Richardson et al. (1968) (squares). The reversal and calculated kyanite-sillimanite boundary slope observed by Newton (1966) are highlighted in red. The symbol dimension encompasses the declared uncertainty on pressure (0.4 kbar; Newton 1966a)

## 6.5.2 *The $Al_2O_3$-$SiO_2$ Binary*

This binary join is quite peculiar because apparently no heat is consumed or released when mixing the two liquid components, and the (quite limited) mixing energy in the amorphous state is purely entropic (and, most likely, configurational). With the heat capacity of Berman (1988) and the H°f, S° of the JANAF tables, the HPA model predicts stoichiometric mullite to melt congruently at T = 2156 K and to form an eutectic with alumina at T = 2153 K and $X_{SiO2}$ = 0.36 (Fig. 6.48). A strain energy contribution to the Gibbs free energy of mixing of the liquid is responsible for the marked depression of the liquidus curve flanking the eutectic point on the silica-rich side of the phase diagram. A deep eutectic with cristobalite at $X_{SiO2}$ = 0.93 and T = 1910 K is thus predicted, nearly similar to what was experimentally determined by Aramaki and Roy (1962) and Schairer and Bowen (1955) (i.e., $X_{SiO2}$ = 0.94 and T = 1914 K). As shown by Eriksson and Pelton (1993), assuming mullite to be non-stoichiometric, the fusion turns out to be incongruent and to take place at a higher T in close agreement with the results of Klug et al. (1987) (T = 2163 K).

Treating mullite as a simple binary mixture is somewhat unorthodox because we must accept that fictive isostructural end-members of the same spatial group do exist. Assuming mullite to be a simple binary mixture to all extents (see Par. 4.4.1) and

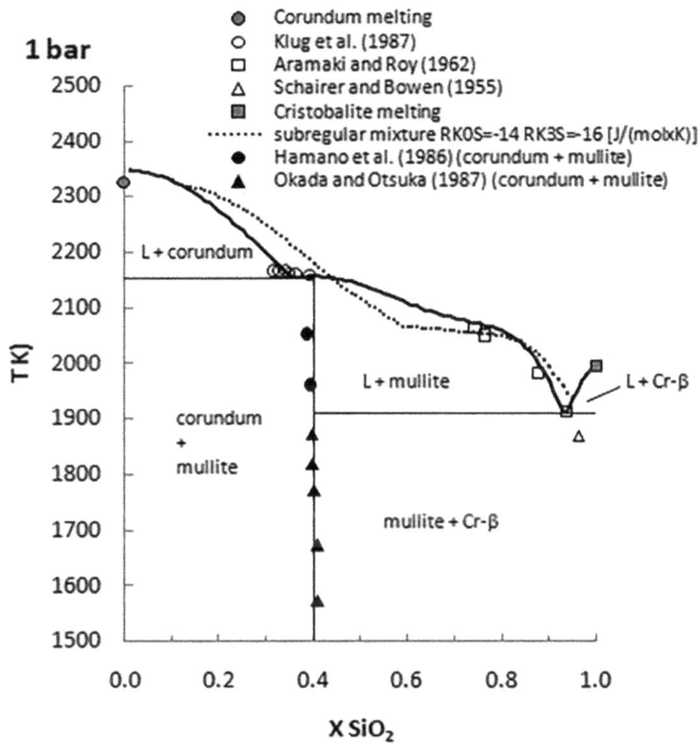


**Fig. 6.48** Liquidus topology of $Al_2O_3$-$SiO_2$ phase diagram at P = 1 bar. HPA model predictions (subregular mixture and stoichiometric compound) are compared with the experimental liquidus observations of Schairer and Bowen (1955), Aramaki and Roy (1962), and Klug et al. (1987), and the subsolidus observations of Hamano et al. (1986) and Okada and Otsuka (1987)

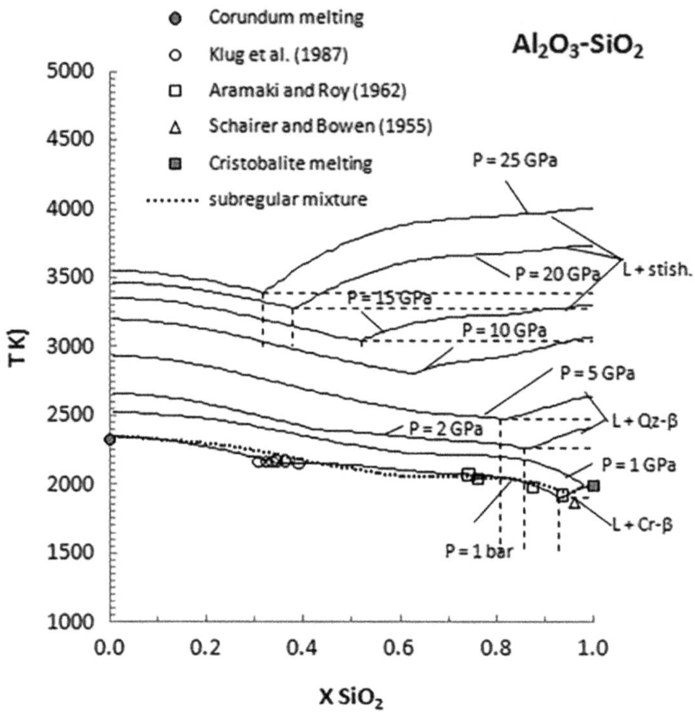


**Fig. 6.49** Liquidus topology of $Al_2O_3$-$SiO_2$ phase diagram from P = 1 bar to P = 25 GPa. Solid lines: stoichiometric mullite. Dotted line: mullite solid mixture

the isostructural end members ($Al_2O_3$ and $SiO_2$) to have the same thermodynamic properties of their stable state at T, P of interest, also HPA predicts it to melt incongruently. The eutectic at high silica content is observed at T = 1939 K and $X_{SiO2}$ = 0.95. We must be conscious, however, that treating mullite as a mixture is probably no more than a computational exercise, as evident in Fig. 6.47, where stoichiometric mullite confirms the subsolidus observations of Hamano et al. (1986) and Okada and Otsuka (1987).

With mullite as a stoichiometric compound, the liquidus temperatures progressively increase with pressure (Fig. 6.49). The eutectic between corundum and $SiO_2$ sensibly moves toward $Al_2O_3$-rich compositions with the increase of P.

## 6.6 CaO-MgO

This join was first investigated by Rankin and Merwin (1916), who found that "*in all fused mixtures of CaO and MgO the two oxides crystallize out side by side, showing no evidence of combination, that is, there is no compound formed stable in contact*

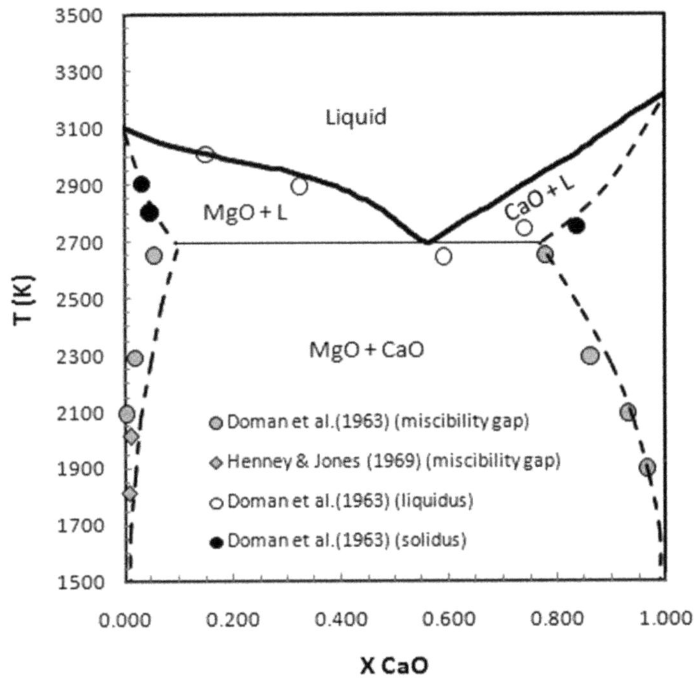


**Fig. 6.50** Roozeboom type-V MgO-CaO phase diagram at P = 1 bar. Experimental observations are superimposed on HPA calculations (solid and dotted lines)

*with melt*". Later investigations showed that a limited mutual solubility does exist, and the join is practically a prototype of Roozeboom type-V phase diagram (i.e., the tangency condition of the two unmixing lobes of the solid mixture finds the Gibbs free energy curve of the liquid in an intermediate position between the unmixing lobes). The rabbit ears are easily reproduced assuming a subregular liquid with a simple Redlich–Kister expansion of just one entropic Van Laar coefficient ($A_{1S} = -30$; see Eqs. 4.33–4.38). The wide subsolidus solvus requires a somewhat more complex treatment of the solid mixture (see Table 4.4). The 1 bar phase diagram so obtained (Fig. 6.50) reproduces satisfactorily the experimental observations of Doman et al. (1963) and Henney and Jones (1969).

# Chapter 7
# The Ternaries

There are four ternary faces limiting the tetrahedron. Besides the solids already seen to nucleate in their limiting binaries, new solid phases form in the inner parts of these systems. For these new solids, we must carry out the same procedures adopted for the other solids and devoted to ascertain thermodynamic and thermophysical consistency and obedience to first principles.

## 7.1 CaO-$Al_2O_3$-$SiO_2$ (CAS)

The new phases forming within CAS are anorthite ($CaAl_2Si_2O_8$), the three gehlenite polymorphs ($Ca_2Al_2SiO_7$), grossular ($Ca_3Al_2Si_3O_{12}$), and Ca-Tschermak ($CaAl_2SiO_6$).

### *7.1.1 $CaAl_2Si_2O_8$ (Anorthite)*

As an end member of the *plagioclase* solid mixture, *anorthite* ($CaAl_2Si_2O_8$) has been the object of extensive investigation. Anorthite is triclinic; its structural class changes, however, from $P1$ symmetry at low-T to $I1$ symmetry at high-T (Bruno et al. 1976; Chiari et al. 1978). It is generally assumed that $CaAl_2Si_2O_8$, based on the *Al-avoidance principle,* has complete ordering of Al and Si on tetrahedral sites at all temperatures. Indeed, some Authors suggest that ordering is not complete at high-T (see, for instance, Bruno et al. 1976). According to Carpenter and MacConnell (1984), the $P1 \Leftrightarrow I1$ transition is a high order. Transition temperatures between the $P1$ and $I1$ forms for pure anorthite at P = 1 bar are estimated by Carpenter and Ferry (1984) to be around 2000–2250 K, and (implicitly) the Authors consider structural transition as coincident with substitutional order–disorder transition. This implicit

© The Author(s), under exclusive license to Springer Nature Switzerland AG 2026 
G. A. Ottonello et al., *Ab-initio Assisted Petrology*,
https://doi.org/10.1007/978-3-032-25658-4_7

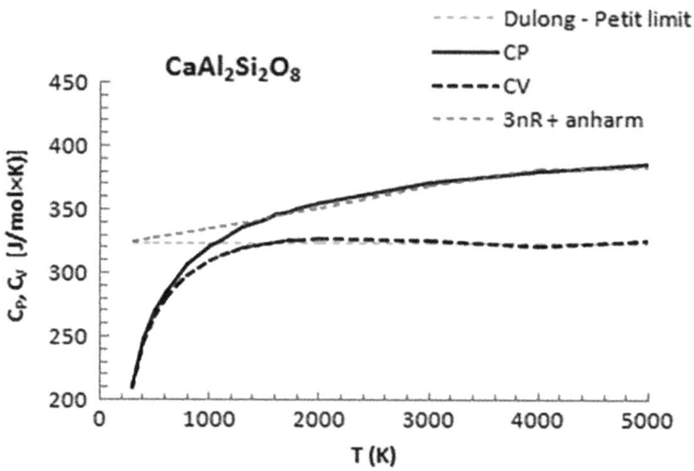


**Fig. 7.1** Isobaric (heavy black solid line) and isochoric (black dashed line) heat capacities of anorthite $C_P$ after Berman (1988), $C_V$ obtained in this study from inverse modeling of anharmonicity

consideration leads to an estimated transition enthalpy of $15.5 \pm 2.5$ kJ/mole and a transition entropy of 5.9–9.2 J/(mole × K). The thermodynamic data available in literature are (though not explicitly) specific to the *P*1 polymorph and do not comprise the transition effect that is, in many cases, outside the T limit of validity of the $C_P = f(T)$ function (Ottonello 1997). Different from a previous investigation (Ottonello et al. 2013), we adopt in this study the $C_P = f(T)$ function of Berman (1988), which proves consistent with first principles in a wide T-range (Fig. 7.1). In the computation of anharmonicity, we adopted the parameterization of Saxena et al. (1993) for thermal expansion and compressibility (this last slightly modified to be extensible to very high-T; see Table 4.2). $H^0_{f,298.15}$, $S^0_{298.15}$, and $V^0_{298.15}$ were initially those of Berman (1988). $H^0_{f,298.15}$ was then lowered by 4 kJ/mol to better reproduce subsolidus equilibria (see later).

### 7.1.2 $CaAl_2SiO_7$ (Gehlenite)

*Gehlenite* is a sorosilicate belonging to the *melilite* (ghelenite–akermanite) series (tetragonal, space group $P42_1m$). Its structural refinement is due to Louisnathan (1969). It formed in the early (hottest) stage of the presolar nebula (carbonaceous chondrites) and exhibits a complex thermal disorder affecting its entropy. The T-dependent disordering model of Berman (1988) "*fits the integrated form of an extended $C_P$function to the enthalpy differences between ordered and disordered phases*" (Berman 1988). Defining as *t* the reference temperature which marks the onset of disordering and $T_D$ the temperature at which the phase is fully disordered, the $C_{P_{ds}}$ contribution is

$$C_{P_{ds}} = d_0 + d_1 T^{-0.5} + d_2 T^{-2} + d_3 T + d_4 T^2 \quad (7.1)$$

Integrating the above equation between $t$ and T ($T < T_D$), entropy and enthalpy effects of disorder are obtained (see Eqs. 16 and 17 in Berman, 1988). Building on this theory, Eriksson and Pelton (1993) computed distinct $C_P$, $H^0_{f,298.15}$, $S^0_{298.15}$ for three gehlenite polymorphs valid in separated T (K)-ranges [i.e., 298–698 (gehlenite-a), 698–1600 (gehlenite-b), >1600 (gehlenite-c)]. Because Berman's procedure cannot be easily exploited in our formalysm (i.e. his Eqs. 15–20; $T_D = 1600$ K, $t = 698$ K, $d_0 = -221.74$, $d_1 = 0$, $d_2 = 1.7291 \times 10^{-7}$, $d_3 = 0.3695$, $d_4 = -1.469 \times 10^4$), we adopt here the reworked data of Eriksson and Pelton (1993), though keeping unvaried the $H^0_{f,298.15}$ of Berman (1988) for the ordered phase (Table 4.2). Although conceived for a limited T range ($298 \leq T \leq 1600$ K), the data of Eriksson and Pelton (1993) may be adopted up to the highest T of interest in our investigation. According to Merlini et al. (2005) the volumetric mean thermal expansion coefficients for gehlenite, obtained with a linear fit of synchrotron X-ray powder diffraction in the temperature range 298–1,400 K, is $2.83 \times 10^{-5}$ $K^{-1}$ and a close look to their data reveal that in this T-range $\alpha$ decreases progressively frm $2.834 \times 10^{-5}$ at 298 K to $2.743 \times 10^{-5}$ at 1473 K. According to the report of Skinner (1966) however $\alpha$ rises progressively from $2.3 \times 10^{-5}$ at T = 293 K to $2.6 \times 10^{-5}$ at T = 1073 K. According to Merlini et al. (2009b) the synchrotron X-ray powder diffraction bulk modulus of gehlenite is 106.4 GPa. We adopt here the thermal expansion of Skinner (1966) ($\alpha_0 = 4.323 \times 10^{-9}$, $\alpha_1 = 2.177 \times 10^{-5}$) and assume that the volume compressibility progressively decreases with the amount of T from 106.4 GPa to attain at the maximum T of interest in this study (5000 K) values consistent with the expected anharmonicity (Fig. 7.2) ($K_0 = 106.4$ GPa, $K_1 = -2.3$ bar/K, $K_2 = -0.0253$ bar/$K^2$; see Table 4.2).

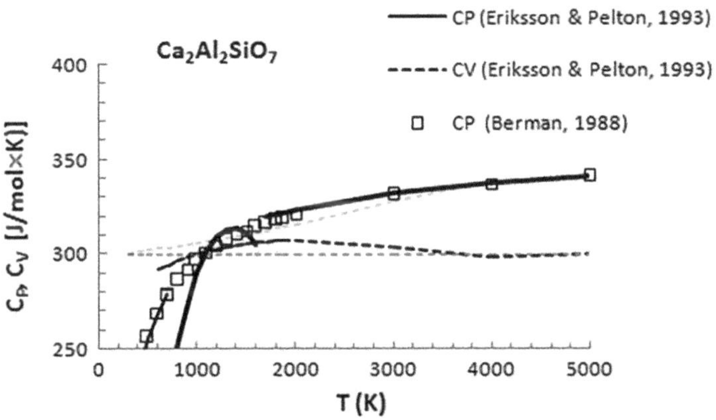


**Fig. 7.2** Isobaric (heavy black solid line) and isochoric (black dashed line) heat capacities of anorthite $C_P$ after Eriksson and Pelton (1993). $C_V$ obtained in this study from inverse modeling of anharmonicity. The red and green dotted lines are, respectively, the Dulong-Petit limit and the expected $C_P$ function at high-T

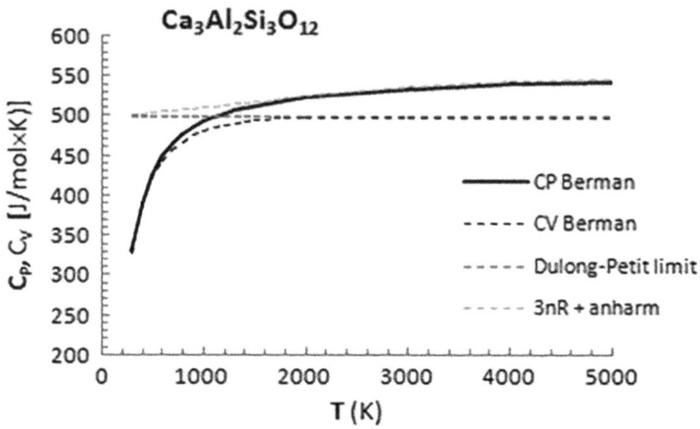


**Fig. 7.3** Isobaric (heavy black solid line) and isochoric (black dashed line) heat capacities of grossular. $C_P$ after Berman (1988). $C_V$ obtained in this study from inverse modeling of anharmonicity. The red and green dotted lines are, respectively, the Dulong-Petit limit and the expected isochoric heat capacity at high-T

### *7.1.3 $Ca_3Al_2Si_3O_{12}$ (Grossular)*

*Grossular* has a body-centered-cubic cell (spatial group *Ia3d*) and eight formula units per unit cell. It belongs to the calcic group of *ugrandites,* comprising also uvarovite ($Ca_3Cr_2Si_3O_{12}$) and andradite ($Ca_3Fe_2Si_3O_{12}$). Its typical occurrence is in contact-metamorphosed limestones (skarns, marbles, cornubianites). As noted by Berman (1988), "*there is considerable disagreement among different sets of measurements of the heat capacity and third law entropy of grossular*". The overall agreement with phase equilibrium data at subsolidus can be obtained only by adopting the T-dependent disordering model of gehlenite. This is basically why, also for grossular, we adopt the thermodynamic parameterization of Berman (1988). The implications of this choice will be elucidated later on. For the thermophysical parameterization, we adopt the values obtained by Ottonello et al. (1996), which are based on standard potentials. The thermal dependence of the bulk modulus is obtained by conforming appropriately the anharmonicity of the substance in order to achieve a reasonable compliance of the isocoric heat capacity with first principles (Fig. 7.3).

### *7.1.4 $CaAl_2SiO_6$ (Calcium-Tschermak)*

*Calcium-Tschermak* is monoclinic (space group *C2/c*), it belongs to the augite series, and is virtually present in solid mixture in all natural clinopyroxeses. According to Okamura et al. (1974) its cell volume is 63.445 cc/mol. Berman (1988) and Saxena et al. (1993) adopted the value of Robie et al. (1978), which is slightly higher (i.e.,

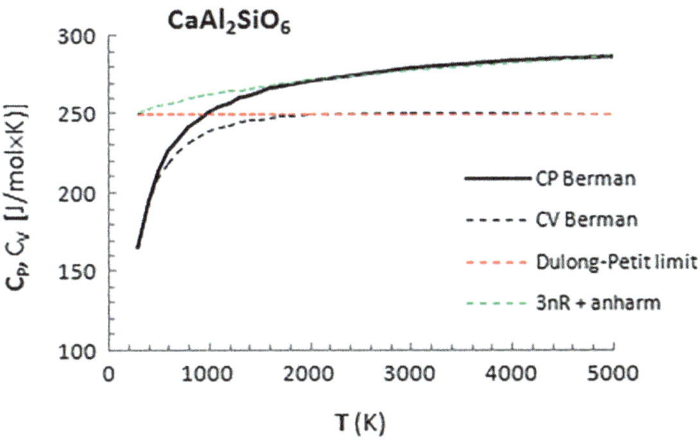


**Fig. 7.4** Isobaric (heavy black solid line) and isochoric (black dashed line) heat capacities of Ca-Tschermak. $C_P$ after Berman (1988), $C_V$ obtained in this study from inverse modeling of anharmonicity, based on the thermophysical parameters of Saxena et al. (1993) (with minor adjustments)

63.56 cc/mol). We adopt here $H^0_{f,298.15} = -3296.8$ kJ/mol and $S^0_{298.15} = 144.5$ J/(mol × K) (Chatterjee 1989; see also Saxena et al. 1993) and for $C_P$ we retain the functional form of Berman (1988). For the standard-state molar volume, we adopt the value of Robie et al. (1978), and for the other thermophysical parameters, we adopt the value proposed by Saxena et al. (1993), with some adjustments of $K_1$, $K_2$ based on the computed anharmonicity (Table 4.2 and Fig. 7.4).

#### 7.1.4.1 Subsolidus Equilibria

Although in this study we are mainly interested in liquidus topology, for its importance in determining the evolution of petrogenetic systems when heat is released, subsolidus equilibria are important as well because they offer an opportunity to check against experiments if the adopted parameterization of thermodynamic and thermophysical data is sufficiently accurate. Berman (1988) carried out an accurate analysis in this sense for the CAS phases. Figure 7.5a basically reproduces his Fig. 6c, with the relative stabilities of the anorthite (An) + gehlenite (Ge) versus grossular garnet (Gr) + corundum (Cor) assemblages. In part B of the same figure, it is shown the relative stability field of Ge + An + wollastonite (Wo) against Gr, and in part C, the relative stability of Wo + An with respect to Gr + quartz (Qz).

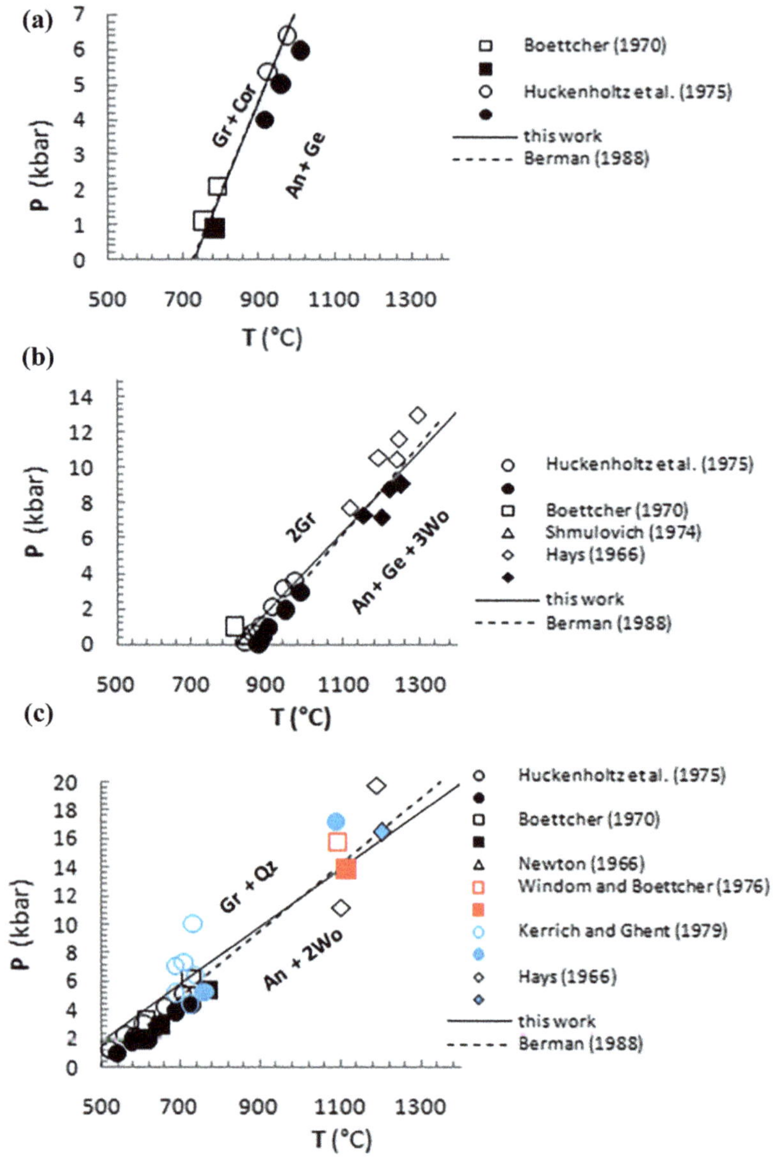


**Fig. 7.5** Computed phase equilibria with respect to experimental observations and the assessments of Berman (1988) in the CAS system. Open + filled symbols are reversals

As noted by Berman (1988), among these phases, perhaps grossular is the most difficult to constrain due to the "*considerable disagreement among different sets of measurements of the heat and third law entropy*". Also, the thermophysical parameters of the phase are affected by huge variability (Ottonello et al. 1996; Ottonello

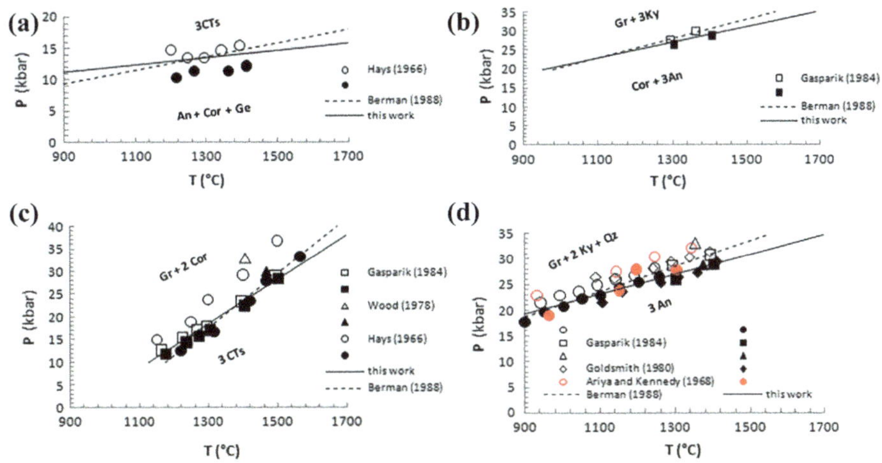


**Fig. 7.6** Computed phase equilibria with respect to experimental observations and the assessments of Berman (1988) in the CAS system. Open + filled symbols are reversals

1997). Adopting for grossular the thermodynamic assessment of Berman (1988) and thermophysical parameters based on first principles, and lowering the original $H^0_{f,298.15}$ of Berman (1988) for An from −4228.73 to −4232.73 kJ/mol, the compliance with experimental observation is sufficiently accurate for our purposes.

Figure 7.6a reproduces Fig. 7c of Berman (1988), with the relative stabilities of the An + Cor + Ge versus Ca-Tschermak (CTs). In part B of the same figure, it is shown the relative stability field of Cor + An versus Gr + kyanite (Ky) [same in Fig. 7b of Berman (1988)]. In part C, it is investigated the relative stability of CTs with respect to the Gr + Cor assemblage [as in Fig. 7d of Berman (1988)] is investigated, and finally, in Fig. 7.3d, is shown the relative stability of An with respect to the Gr + Ky + Qz assemblage [as in Fig. 7a of Berman (1988)]. As anticipated, the limits involving An were obtained by lowering the enthalpy of formation from the elements of 4 kJ/mol with respect to the value adopted by Berman (1988). The results of this exercise are quite satisfactory (Figs. 7.5 and 7.6), and we may now confidently move to the construction of the CAS topology at liquidus.

#### 7.1.4.2 CAS Topology at Liquidus

Figure 7.7 shows a 3D representation of the liquidus surface of the CAS system at P = 1 bar conditions. The surface was obtained by application of the convex-hull procedure, which determines a set of sample points interpolated through a lifted Delaunay

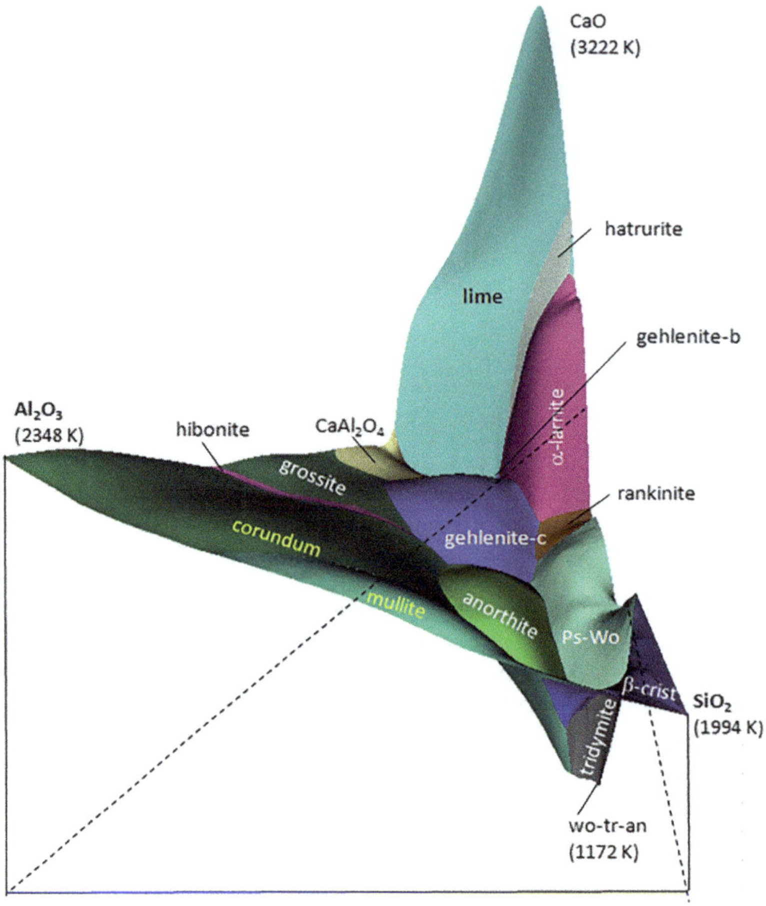


**Fig. 7.7** The liquidus surface of CAS at P = 1 bar. Step size = 5 K; compositional increments = 0.005. Smoothing runs = 20

triangulation. Calculations were performed from super-liquidus to subsolidus conditions (3000–1000 °C) using a step size of 5 K and a discretized grid of $4 \times 10^6$ points per temperature. All solids are treated as mechanical mixtures (mullite solid mixing has a negligible effect on phase topology at liquidus, and is here neglected for the reasons previously outlined at Sect. 4.4.1). The present study basically reproduces what was already seen by Ottonello et al. (1993), though the ternary interactions have been here substantially simplified. We have a limited loss in precision (both in T and position), but the model is apparently more robust, especially at high P.

The T range encompassed by the liquidus hypersurface is more than 2000 K, and the steepest surfaces are observed near the CaO-$SiO_2$ join. In passing, we may note that the fundamental difference with respect to our previous HPA optimization (Ottonello et al. 2013; see their Fig. 7.3) arises mainly from the fact that CaO is expected here to melt at a much higher T (i.e., more than 350 K of difference). As anticipated, also the way of treating the ternary interaction is here different: instead of mimicking the effect of CaO in determining the periodicity of the Al-Si chains through a high-order Redlich–Kister expansions (all in all five parameters;

**Table 7.1** Calculated HPA (3) versus experimental ternary invariants (1) and MQC assessment (2). (1): Osborn and Muan 1966); (2) Eriksson and Pelton (1993); (3): this work. The mean absolute bias on temperatures between HPA and experients is 65.4 K

| Solid phases | | | Type | T(k) | Molar proportions | | | Refs. |
|---|---|---|---|---|---|---|---|---|
| a | b | c | | | CaO | $Al_2O_3$ | $SiO_2$ | |
| Anorthite | | | Congruent | 1828 | 0.250 | 0.250 | 0.500 | 1 |
| | | | | 1828 | 0.250 | 0.250 | 0.500 | 2 |
| | | | | 1827 | 0.250 | 0.250 | 0.500 | 3 |
| Gehlenite (c) | | | Congruent | 1868 | 0.500 | 0.250 | 0.250 | 1 |
| | | | | 1868 | 0.500 | 0.250 | 0.250 | 2 |
| | | | | 1866 | 0.500 | 0.250 | 0.250 | 3 |
| $CaAl_2O_4$ | Gehlenite (b) | Bredrigite | Peritectic | 1655 | 0.598 | 0.290 | 0.113 | 1 |
| | | | | 1734 | 0.600 | 0.250 | 0.150 | 2 |
| | | | | 1622 | 0.652 | 0.177 | 0.171 | 3 |
| Grossite | $CaAl_2O_4$ | Gehlenite (c) | Eutectic | 1775 | 0.497 | 0.387 | 0.116 | 1 |
| | | | | 1802 | 0.490 | 0.430 | 0.080 | 2 |
| | | | | 1776 | 0.545 | 0.337 | 0.118 | 3 |
| High-tridymite | Anorthite | Mullite | Eutectic | 1620 | 0.115 | 0.127 | 0.758 | 1 |
| | | | | 1640 | 0.117 | 0.132 | 0.750 | 2 |
| | | | | 1522 | 0.104 | 0.128 | 0.768 | 3 |
| Gehlenite (c) | Larnite | Rankinite | Peritectic | 1590 | 0.531 | 0.070 | 0.398 | 1 |
| | | | | 1601 | 0.517 | 0.085 | 0.398 | 2 |
| | | | | 1743 | 0.532 | 0.098 | 0.370 | 3 |
| Ps-wollastonite | Rankinite | Gehlenite (c) | Peritectic | 1585 | 0.515 | 0.072 | 0.413 | 1 |
| | | | | 1601 | 0.485 | 0.096 | 0.419 | 2 |
| | | | | 1673 | 0.461 | 0.118 | 0.421 | 3 |
| Ps-wollastonite | Anorthite | Gehlenite (c) | Eutectic | 1540 | 0.430 | 0.125 | 0.445 | 1 |
| | | | | 1537 | 0.449 | 0.114 | 0.436 | 2 |
| | | | | 1646 | 0.424 | 0.136 | 0.439 | 3 |

(continued)

**Table 7.1** (continued)

| Solid phases | | | Type | T(k) | Molar proportions | | | Refs. |
|---|---|---|---|---|---|---|---|---|
| a | b | c | | | CaO | $Al_2O_3$ | $SiO_2$ | |
| Wollastonite | High-tridymite | Anorthite | Eutectic | 1445 | 0.259 | 0.093 | 0.648 | 1 |
| | | | | 1463 | 0.276 | 0.083 | 0.641 | 2 |
| | | | | 1371 | 0.166 | 0.072 | 0.762 | 3 |
| Gehlenite (c) | Hibonite | Grossite | Peritectic | 1745 | 0.398 | 0.313 | 0.288 | 1 |
| | | | | 1716 | 0.386 | 0.284 | 0.331 | 2 |
| | | | | 1785 | 0.391 | 0.336 | 0.272 | 3 |
| Anorthite | Mullite | Corundum | Peritectic | 1787 | 0.193 | 0.251 | 0.556 | 1 |
| | | | | 1820 | 0.226 | 0.297 | 0.477 | 2 |
| | | | | 1765 | 0.168 | 0.245 | 0.587 | 3 |
| Larnite | Hatrurite | Lime | Peritectic | 2073 | 0.700 | 0.070 | 0.230 | 2 |
| Larnite | Gehlenite | Bredrigite | Peritectic | 1705 | 0.532 | 0.098 | 0.370 | 3 |
| Larnite | Lime | Bredrigite | Eutectic | 1737 | 0.655 | 0.151 | 0.194 | 3 |
| Corundum | Anorthite | Gehlenite | Peritectic | 1730 | 0.364 | 0.285 | 0.351 | 3 |
| Corundum | Hibonite | Gehlenite | Peritectic | 1777 | 0.385 | 0.328 | 0.287 | 3 |

see Eqs. 13–18 in Ottonello et al. 2013), here we use a simple ternary coefficient (see Sect. 4.5). The price to pay for this obedience to Occam's Razor is acceptable, both in terms of T and composition (Table 7.1). All in all, the ternary is resolved with 13 parameters (4 parameters for the chemical energy, 1 parameter for polarization, and eight parameters for strain energy contributions; see Eq. 4.35 and Sect. 4.5). The mathematical elasticity is thus almost at the MQC modeling level (Eriksson and Pelton 1993), i.e., 16 parameters (14 parameters for the binary interactions and two parameters for the ternary contributions).

In terms of energy, we may evaluate the model accuracy by imposing the stoichiometry and the temperature of the (experimental) invariant point and calculating the actual amount of energy necessary to return a perfect energy balance. The absolute difference with the HPA model energy, for the same T-x conditions, allows us to establish the Mean Absolute Deviation (MAD; Fig. 7.8). The MAD obtained on the invariants of Osborn and Muan (1966) is 2.118 kJ/mol. The maximum scatter in the experimental values is observed at $X_{SiO2} \approx 0.4$ (Fig. 7.8).

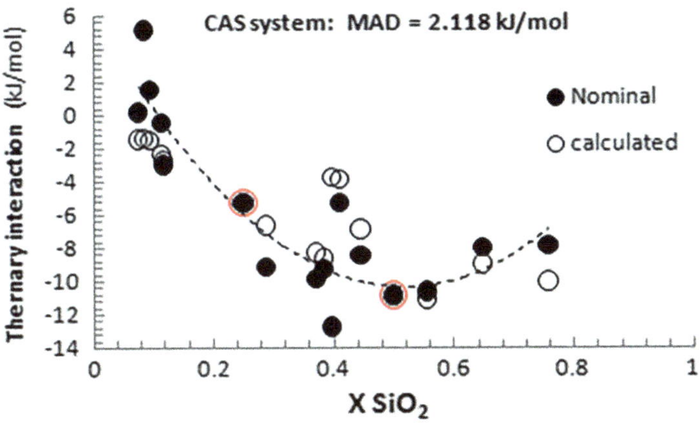


**Fig. 7.8** Ternary excess energy in the CAS system at P = 1 bar. Nominal values are compared with calculations. The red-circled nominal invariants are the congruent melting points of anorthite and gehlenite, adopted to constrain the polarization energy (see Sect. 4.5)

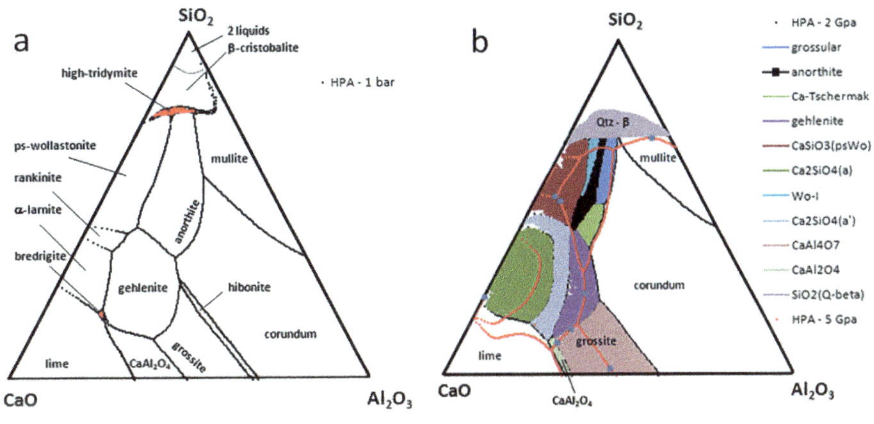


**Fig. 7.9** Primary phase boundaries of the CAS system at 1 bar (**a**) and 2 and 5 GPa (**b**)

Considering that the errors introduced in calculations by a wrong $C_P = f(T)$ function for the nucleating solids may be more than two orders of magnitude higher, we consider the model performance acceptable and move thus to the evaluation of P effects on the system. These effects are illustrated by Schreinemaker's projection of the liquidus boundaries on the CAS compositional plane (Fig. 7.9).

Increasing P from room conditions to 2 GPa, many changes in the primary phase fields are observed, namely:

- β-quartz replaces β-cristobalite and the limited field of high-tridymite;
- The field of anorthite is predated by grossular, Ca-Tschermak, and wollastonite-I;
- The fields of mullite and pseudo-wollastonite shrink;

- The field of bredrigite expands and the field of gehlenite if predated by α-larnite;
- The field of corundum expands considerably predating hibonite;
- The field of grossite expands predating $CaAl_2O_4$.;
- Rankinite is no longer present as the primary phase at the liquidus.

The observation that at P = 2 GPa grossular and Ca-Tschermak predate the field of anorthite is in line with the subsolidus observations of Hays (1966), Hariya and Kennedy (1968), Wood (1978), Goldsmith (1980), and Gasparik (1984). Grossular completely replaces anorthite at P = 5 GPa and is limited univariantly by coesite at high $SiO_2$ content. The stability field of mullite progressively shrinks with pressure, and P = 5 GPa is no longer present as the primary phase at the liquidus.

In a instructive discussion elucidating the complex nature of the peritectic melting of anorthite at high P [kept in evidence by the experiment of Hariya and Kennedy (1968)], Morse (1980) anticipated that this is due to the fact that the eutectic between Corundum and Anorthite, which, at P = 1 bar occurs at a composition relatively richer in $Al_2O_3$ component along the $X_{CaO} = 0.25$ pseudo-binary, is progressively displaced at lower $Al_2O_3$ amounts with the increase of P attaining the stoichiometric $X_{Al2O3} / (X_{Al2O3} + X_{SiO2}) = 1/3$ ratio of anorthite at P ≈ 9 kbar and trenspassing it at higher P (see Fig. 18.4 in Morse 1980). P = 9 kbar is thus a peculiar condition at which the *cotectic* curve transforms into a *peritectic* one. In Fig. 7.10, we see that our calculations confirm this smart anticipation. This also indicates that excess high-order volume terms may be safely assumed to be absent in the liquid and that our simplicial surfaces at high P are sufficiently accurate representations of the actual phase topology of the system at liquidus.

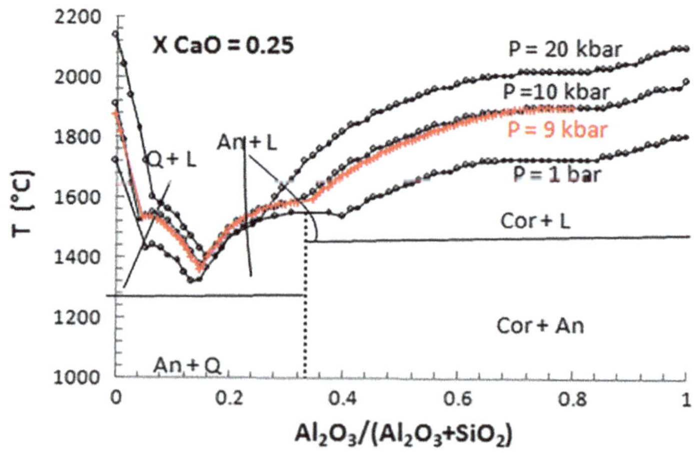


**Fig. 7.10** The anorthite pseudo-binary at various P conditions. The eutectic between Corundum (Cor) and Anorthite (An) moves progressively leftward with the increase of P and attains the Anorthite ratio 1/3 at P ≈ 9 kbar (red solid line). The liquidus curves are resolved by the improved convex-hull procedure and correspond to transetcs of CAS at $X_{CaO} = 0.25$

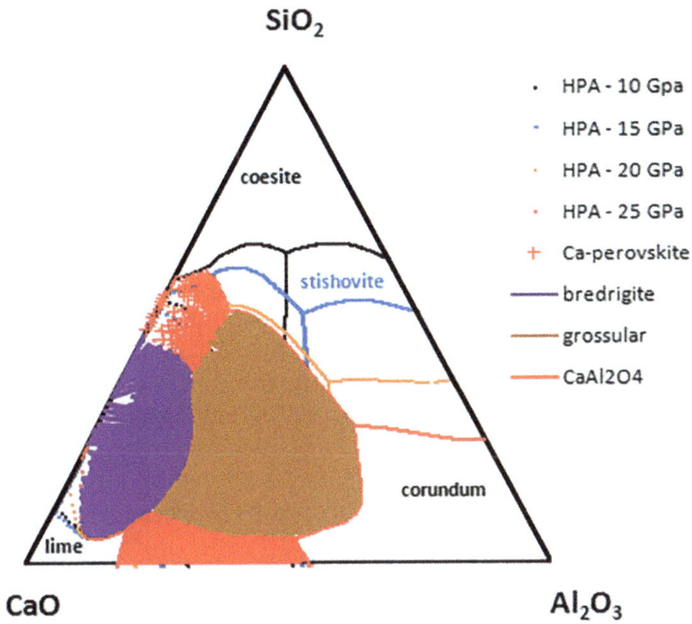


**Fig. 7.11** Primary phase boundaries of the CAS system at various pressures from 10 to 25 GPa

We see, finally, in Fig. 7.11, the topological behavior of CAS at very high pressures.

At P = 10 GPa, the stable silica polymorph is coesite, which is replaced by stishovite at 15 GPa. At P = 25 GPa, Ca-Perovskite is present as the primary phase at liquidus, and $CaAl_2O_4$ is present again, predating all the other Ca-aluminates. The stability field of stishovite and grossular progressively occupies larger portions of the topological space as pressure increases. The primary phase field of corundum shrinks considerably with the increase of P.

## 7.2 MgO-$Al_2O_3$-$SiO_2$ (MAS)

This system was recently investigated by Belmonte et al. (2017b), and the present assessment does not change much of what was previously observed. As in the case of CAS, also in the MgO-$Al_2O_3$-$SiO_2$ system, 3-components solids form in the inner part of the system. The new phases forming within MAS are sapphirine-351 ($Mg_3Al_{10}SiO_{20}$), sapphirine-442 ($Mg_4Al_8Si_2O_{20}$), cordierite ($Mg_2Al_4Si_5O_{18}$), and pyrope ($Mg_3Al_2Si_3O_{12}$).

### 7.2.1 Sapphirines

Sapphirine is a chain silicate with simplified general formula $M_8T_6O_{20}$, where M and T represent octahedral and tetrahedral sites, respectively (Deer et al. 1997). Two different structural modifications, related to the Orde-Disorder (OD) character of the mineral (Merlino 1973), have been proposed in the literature: a monoclinic 2M polytype with eight independent octahedral sites (M1, M2, M3, M4, M5, M6, M7, and M8) and six independent tetrahedral sites (T1, T2, T3, T4, T5, and T6) (Moore 1969); a triclinic 1A polytype with a different stacking sequence and an additional independent octahedral site (M9) (Merlino 1980). A further structural complexity is represented by the degree of cation ordering on M and T sites, for which there is no general consensus among different structural determinations (Moore 1969; Higgins and Ribbe 1979; Merlino 1980; Christy et al. 1992). If only $MgO$-$Al_2O_3$-$SiO_2$ compositions are taken into account, Mg and Al atoms are distributed over octahedral sites, while Al and Si atoms occupy tetrahedral sites. Compositional variations may occur along the binary join according to the Tschermak-type coupled substitution:

$$^{M}Mg^{2+} + {}^{T}Si^{4+} \Leftrightarrow {}^{M}Al^{3+} + {}^{T}Al^{3+} \tag{7.2}$$

The 2M polytype of sapphirine has a monoclinic structure, space group *P21/a*, with 136 atoms in the primitive unit cell (Z = 4 for the 20-oxygen formula unit), all in the general position. This polytype has been the object of detailed *ab-initio* calculations. Using the hybrid B3LYP density functional method, Belmonte et al. (2014) computed the ab-initio thermodynamic and thermophysical properties of the two sapphirine end-members, $Mg_4Al_8Si_2O_{20}$ (sapphirine-442) and $Mg_3Al_{10}SiO_{20}$ (sapphirine-351), in the join $Mg_3(Mg_{1-X}Al_X)Al_8(Al_XSi_{1-X})SiO_{20}$ with X = 0 to 1. Static and vibrational calculations, performed in the framework of the quasi-harmonic approximation, allowed us to define the equation of state (EoS), elastic constant tensor, seismic velocities, IR spectra, mode Grüneisen parameters, and thermodynamic properties of both sapphirine end-members. Their B3LYP volumes are slightly overestimated with respect to experiments due to the well-known effect of exchange–correlation functionals based on the generalized gradient approximation (GGA; see Table 9 in Belmonte et al. 2014). On the other hand, the inclusion of some fraction (20%) of non-local exact Hartree–Fock exchange in the hybrid density functional improved the quality of the calculated vibrational spectra, hence the thermodynamic accuracy (for almost insulating crystalline solids; see Ottonello 2024a). Fitting the ab-initio pressure–volume curves with a third-order finite strain Birch-Murnaghan equation of state (BM3-EoS) yielded a static bulk modulus $K_0 = 158.7 \pm 0.6$ GPa and a pressure derivative $K' = 4.68 \pm 0.14$ for sapphirine-442, $K_0 = 172.3 \pm 0.4$ GPa and $K' = 4.28 \pm 0.09$ for sapphirine-351. A modified Kieffer's model was adopted to evaluate the optic and acoustic mode contributions to thermodynamic functions stemming from ab-initio phonon frequencies and directionally averaged seismic velocities, respectively. The ab-initio assessment of the enthalpy of formation from the elements of the two polytypes is somewhat cryptic due to the difficulty in estimating appropriate electronic energies for the isolated gaseous atoms to adopt in thermochemical cycle calculations (see Eqs. 24–26 and Table 9 in Belmonte et al. 2014; see also Ottonello 2024a). The ab-initio value of sapphirine-442 obtained by Belmonte et al. (2014) is −11,158 kJ/mol against −11,005 ± 33.5 kJ/mol measured by Kiseleva (1976) and the value calculated for sapphirine-351 is $H^0{}_f = -11{,}335.3$ kJ/mol against −11,106 ± 8. Both experiments and calculus tell us that the 442 polytype is significantly less stable than 351. In this work, we adopt for the 442 polytype the experimental value of Kiseleva (1976), while for the polytype 351 we adopt −11,199.3 kJ/mol, as in Belmonte et al. (2017). We refined the ab-initio heat capacity functions of the two polytypes (Belmonte et al. 2014) in order to achieve fulfillment of first principles in the wider T range of interest in this study. The refined isobaric and isochoric heat capacities of the two polytypes are shown in Fig. 7.12 (see also Tables 4.1 and 4.4).

### 7.2.2 *Cordierite*

Cordierite is an orthorhombic pseudo-hexagonal cyclosilicate (space group *Cccm*; Z = 4) occurring typically in regional or contact metamorphism of pelitic rocks (in association with sillimanite, corundum, spinel, garnet, andalusite). Because it accepts a variable amount of $OH_2$ in the structure (Mirwald 1982), it has been the object of extensive investigation by petrologists (see references in Berman 1988). Here, we are concerned only with the anhydrous phase. We adopted initially for $H^0{}_f$, $V^0$, $S^0$, and $C_P$ the assessment of Berman (1988) (Fig. 7.13) and for $K_0$, $K_1$ the indications of Haussühl et al. (2011) with $K'$ fixed to 4.0. Thermal expansion,

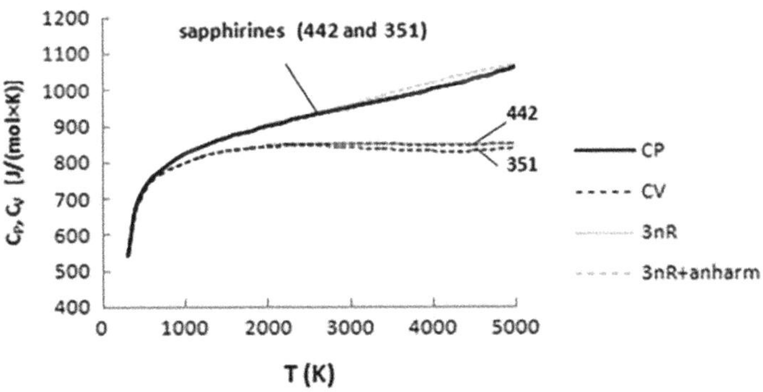


**Fig. 7.12** isobaric (heavy black solid line) and isochoric (black dashed line) heat capacities of sapphirines

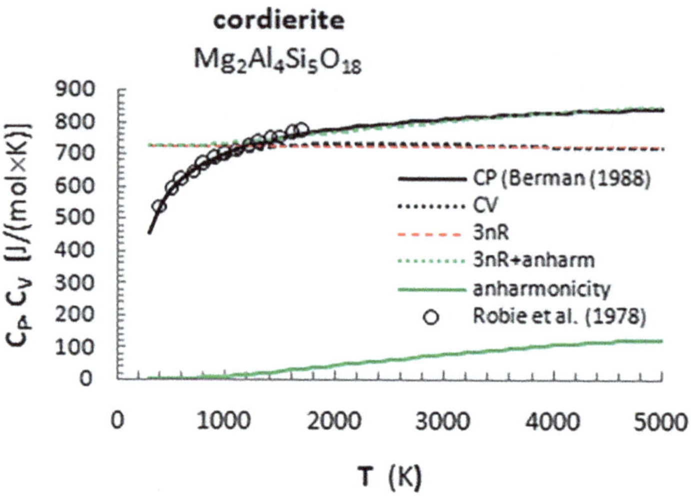


**Fig. 7.13** Isobaric heat capacity (heavy black solid line and empty circles) and isochoric counterpart (black dashed line) of anhydrous cordierite. The Dulong-Petit limit and the anharmonicity are also plotted

initially from Haussühl et al. (2011) was remodeled in order to achieve an isochoric thermal expansion at high-T in line with first principles. $H^0{}_f$ and $S^0$ were remodeled in order to reproduce the observed subsolidus equilibria (Fig. 7.13, 7.14; Tables 4.1 and 4.2).

### *7.2.3 $Mg_3Al_2Si_3O_{12}$ (Pyrope)*

*Pyrope*, as grossular, has a body-centered-cubic cell (spatial group $Ia\bar{3}d$) and eight formula units per unit cell. It is often found in contact metamorphosed limestones. As done in our previous investigation of MAS, we adopt here for the isobaric heat capacity the data of Akaogi et al. (2002). $V^0{}_{298.15}$ is from Berman (1988); $K_0$ and K' are from Erba et al. (2014); $(dK/dT)_P$ from Anderson et al. (1991); $\alpha_V$ from Suzuki and Anderson (1983) (Tables 4.1, 4.2 and 4.4). As we may see in Fig. 7.15, the ensuing anharmonicity fulfills perfectly first principles, and the isochoric heat capacity attains the Dulong-Petit limit at high-T, as expected from theory.

$S^0{}_{298.15}$ and $H^0{}_{f,298.15}$, initially from Berman (1988), were refined, based on the subsolidus equilibrium

$$Mg_2SiO_4 + Mg_3Al_2Si_3O_{12} \Leftrightarrow MgAl_2O_4 + 4MgSiO_3 \quad (7.3)$$

(Perkins et al. 1981; Danckwerth and Newton 1978; Gasparik and Newton 1984). $H^0{}_{f,298.15}$ was reduced by roughly 11 kJ/mol (from −6286.5 to −6297.7 kJ/mol) and

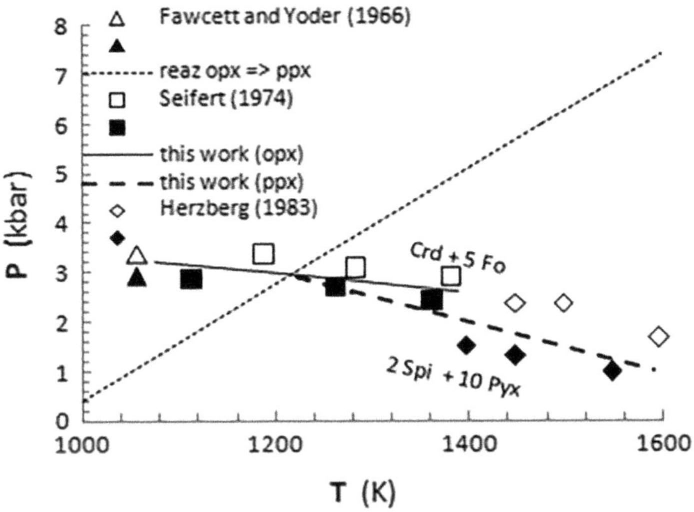


**Fig. 7.14** Computed subsolidus phase equilibria involving cordierite, with respect to experimental observations in the MAS system. Open + filled symbols are reversals. opx = orthopyroxene, ppx = protopyroxene; Pyx = pyroxenes, Crd = cordierite, Spi = spinel, Fo = forsterite

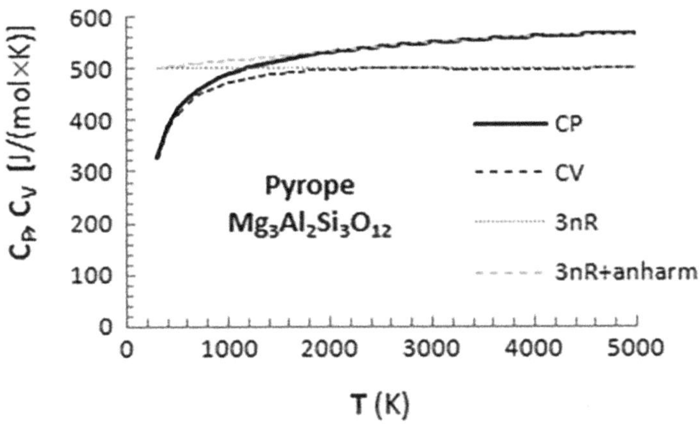


**Fig. 7.15** Isobaric heat capacity (heavy black solid line) and isochoric counterpart (black dashed line) of pyrope

$S^0_{298.15}$ was also reduced by roughly 12 J/(mol × K), [from 266.359 J/(mol × K) to 254.479 J/(mol × K; Table 41]. The new enthalpy is nearer to the calorimetric datum of Charlu et al. (1975) (i.e., −6291.5 kJ/mol). Phase assemblages determined by the above equlibrium (Eq. 7.3) are delineated in Fig. 7.16.

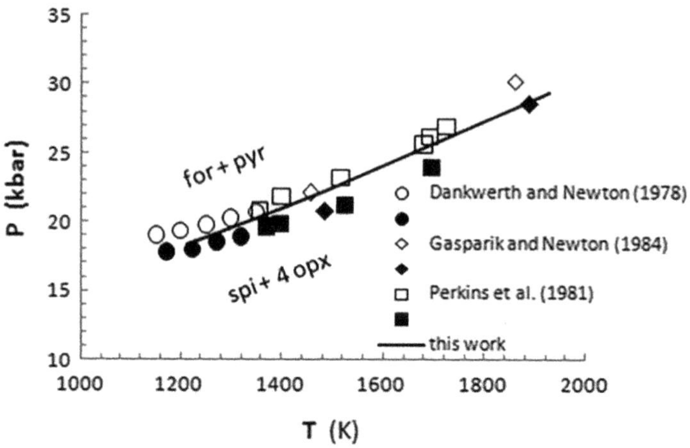


**Fig. 7.16** computed phase equilibrium (Eq. 7.3) with respect to experimental observations. Open + filled symbols are reversals

#### 7.2.3.1 MAS Topology at Liquidus

Figure 7.17 is a 3D representation of the liquidus surface of the MAS system at P = 1 bar conditions. Calculations were performed as already outlined for CAS. We begin treating all nucleating solids as mechanical mixtures. The present study basically reproduces what was already seen by Belmonte et al. (2017b), though the ternary interaction has been simplified here again.

The T range encompassed by the liquidus hypersurface is roughly 1500 K, and the steepest surfaces are observed, as already seen for CAS, along the NM-NF join. All in all, the ternary is resolved with 14 parameters (5 parameters for the chemical energy, 1 parameter for polarization, and five parameters for strain energy contributions; see Eq. 4.35 and Sect. 4.5). In Table 7.2, the HPA ternary invariant points are compared with the results of the MQC assessment (Jung et al. 2004) and experimental observations.

The MAD between calculus and experiments is rather low (1.275 kJ/mol) and, as already seen for CAS, the maximum scatter is observed at intermediate silica content (Fig. 7.18). Moreover, not all experimentally observed invariants are reproduced. Nevertheless, based on topology, the HPA model performance is still acceptable, and we confidently proceed to evaluate the P effects on the system.

In analogy to what has already been seen for CAS, the effects of P on the relative extensions of the primary phase stability fields are illustrated by Schreinemaker's projection of the liquidus boundaries on the MAS compositional plane (Fig. 7.19).

With the increase of P, the primary phase fields of cordierite and sapphyrine shrink progressively. Sapphyrine is no longer present at P = 1 GPa, and cordierite is absent at P = 2 GPa. Also, the field of mullite progressively shrinks with pressure, predated by coesite and corundum, which expand with pressure. The primary phase field of

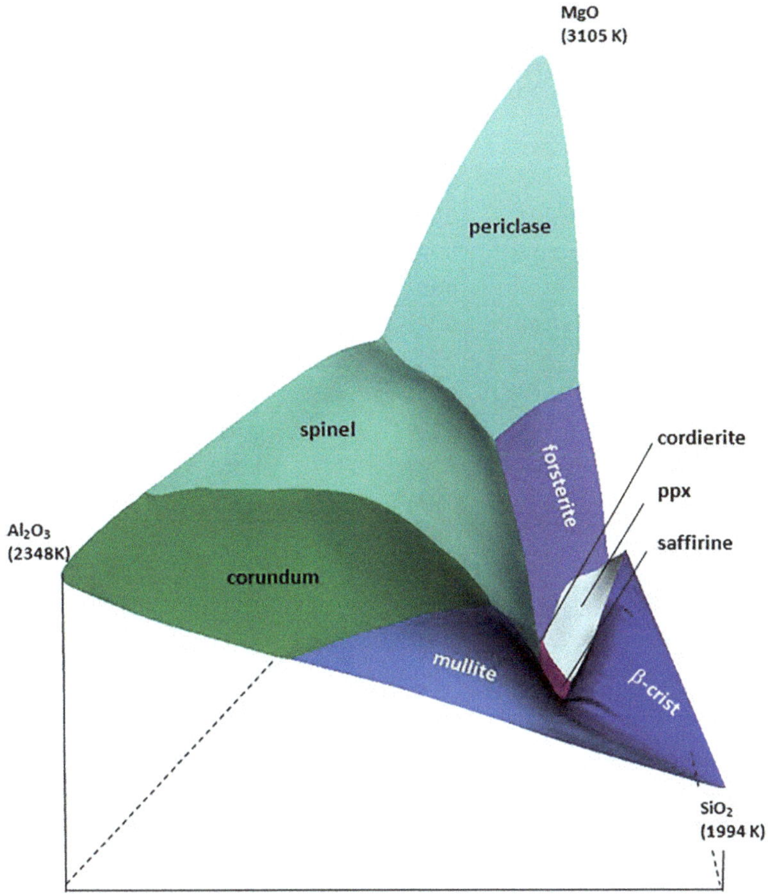


**Fig. 7.17** The liquidus surface of MAS at P = 1 bar. Step size = 5 K; compositional increments = 0.005. Smoothing runs = 20

periclase is virtually unaffected by pressure. The field of forsterite is progressively predated by pyroxene. The field of spinel is progressively predated by corundum.

We see finally in Fig. 7.20 the topological behavior of MAS at very high pressures. At P = 10 GPa, the stable silica polymorph is coesite. The other stable phases at liquidus are forsterite, pyrope, and spinel. At P = 20, majorite predates part of the primary phase field of pyrope. At 25 GPa, perovskite replaces majorite and further predates the pyrope primary phase field. The stable solid, besides stishovite, periclase, and corundum, is only pyrope. The primary phase field of periclase is virtually unaffected by P at all pressures, while the primary phase field of corundum progressively shrinks.

**Table 7.2** Calculated HPA (1) versus MQC (2). Assessments and experimental thernary invariants: (3): Rankin and Merwin (1918); (4) Schreyer and Schairer (1961a, b); (5): Keith and Schairer (1952); (6): Smart and Glasser (1976); (7): Foster (1950)

| Phases | | | MgO | $Al_2O_3$ | $SiO_2$ | T(K) | Refs. |
|---|---|---|---|---|---|---|---|
| Cr-beta | Mull | H-trid (per) | 0.173 | 0.168 | 0.659 | 1720 | 1 |
| | | | 0.146 | 0.144 | 0.710 | 1738 | 2 |
| Crd | Mull | H-trid (eu) | 0.191 | 0.161 | 0.648 | 1642 | 1 |
| H-trid | Px | Cr-beta (per) | 0.303 | 0.073 | 0.624 | 1745 | 1 |
| | | | 0.402 | 0.033 | 0.565 | 1738 | 2 |
| Crd | Px | H-trid (eu) | 0.242 | 0.126 | 0.633 | 1602 | 1 |
| | | | 0.338 | 0.081 | 0.581 | 1634 | 2 |
| | | | 0.295 | 0.105 | 0.599 | 1618 ± 5 | 3 |
| | | | 0.297 | 0.100 | 0.603 | 1629 | 4 |
| Crd | Sapph | Mull (per) | 0.258 | 0.177 | 0.565 | 1646 | 1 |
| | | | 0.294 | 0.200 | 0.506 | 1727 | 2 |
| | | | 0.259 | 0.216 | 0.525 | 1733 ± 5 | 5 |
| | | | | | | 1727 ± 2 | 6 |
| Crd | Fo | Px (eu) | 0.302 | 0.149 | 0.549 | 1621 | 1 |
| | | | 0.397 | 0.092 | 0.512 | 1644 | 2 |
| | | | 0.360 | 0.119 | 0.521 | 1633 ± 5 | 3 |
| | | | 0.357 | 0.115 | 0.528 | 1637 ± 2 | 4 |
| | | | | | | 1633 ± 2 | 6 |
| Fo | Sapph | Crd (per) | 0.301 | 0.156 | 0.542 | 1623 | 1 |
| Sapph | Sp | Fo (per) | 0.310 | 0.162 | 0.528 | 1639 | 1 |
| Sapph | Mull | Sp (per) | 0.282 | 0.232 | 0.486 | 1795 | 1 |
| | | | 0.297 | 0.208 | 0.495 | 1738 | 2 |
| | | | 0.270 | 0.233 | 0.497 | 1755 ± 2 | 5 |
| | | | | | | 1748 | 7 |
| | | | | | | 1742 ± 2 | 6 |
| Mull | Crn | Sp (per) | 0.277 | 0.257 | 0.466 | 1850 | 1 |
| | | | 0.270 | 0.308 | 0.422 | 1848 | 2 |
| | | | | | | 1833 ± 15 | 6 |
| Per | Fo | Sp (per) | 0.619 | 0.148 | 0.233 | 1996 | 1 |
| | | | 0.593 | 0.076 | 0.330 | 1918 | 2 |
| | | | 0.690 | 0.078 | 0.232 | 1973 ± 25 | 5 |

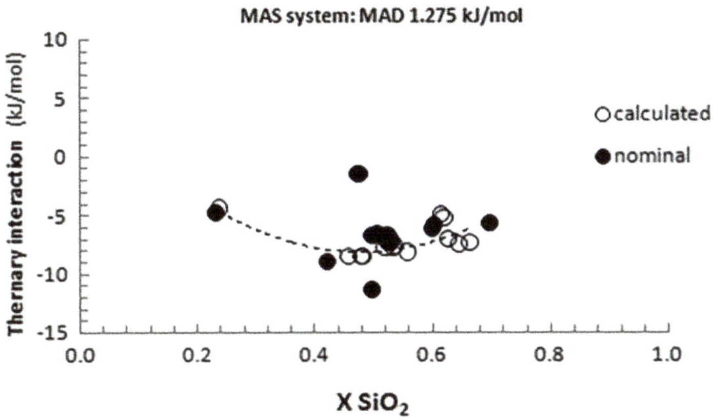


**Fig. 7.18** Ternary excess Energy in the MAS system at P = 1 bar. Nominal values are compared with calculations

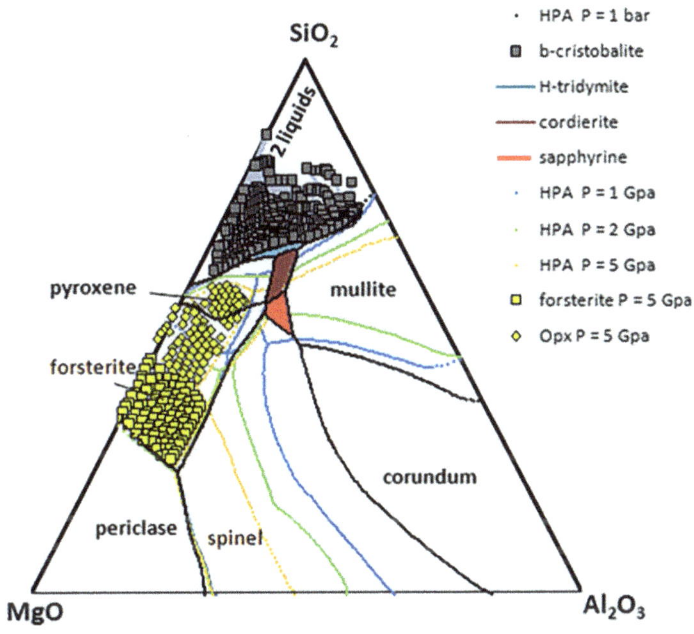


**Fig. 7.19** Primary phase boundaries of the MAS system at various pressures from 1 bar to 5 GPa

## *7.2.4 Topological Effects of Solid-State Mixing*

Within the MAS system, six solid phases exhibit solid-state mixing capability: mullite, pyrope, orthopyroxene, perovskite, spinel, and sapphirine. We have already seen that mullite mixing is quite unlikely for simple thermodynamic reasons, and

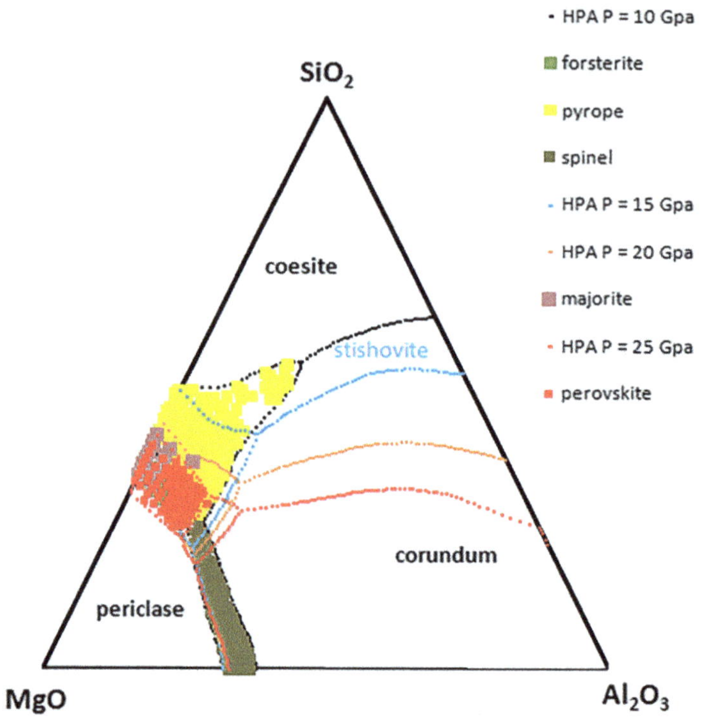


**Fig. 7.20** Primary phase boundaries of the MAS system at various pressures from 10 to 25 GPa

pyrope and perovskite do not appear as phases at the liquidus at low pressure. Only three remaining solid phases could be expected to affect the MAS topology at P = 1 bar, but the Delaunay fragmentation and convex-hull analysis of equilibrium loci tell us that this effect is quite limited (Fig. 7.21). The only sensible effect could be observed only in the central part of the diagram when assuming ideal mixing between the two sapphirines end members (a limited expansion of the primary phase field counterbalanced by an expansion of the multiple oxides field). This expansion also affects the primary phase field of corundum.

In Fig. 7.22, we see the topology of MAS at P = 9 kbar, the pressure condition at which most likely MORB are generated by fusion at the cusp (Presnall et al.1979). Again, solid-state mixing does not alter the position of invariants, but chemical zoning of orthopyroxene is clearly delineated with a progressive increase of $Al_2O_3$ in the mixture moving toward the $Al_2O_3$ apex. Protopyroxene is the stable polymorph at low $Al_2O_3$ content in the system. The true eutectic of the system is at T = 1749 K, where an Opx mixture with 10% moles of $AlAlO_3$ coprecipitates with β-quartz and $Al_6Si_2O_{13}$.

The MAS liquidus surface calculated at P = 23 GPa displays a rather wide primary phase field of majorite-pyrope garnet in a zone of relatively low temperature (Fig. 7.23).

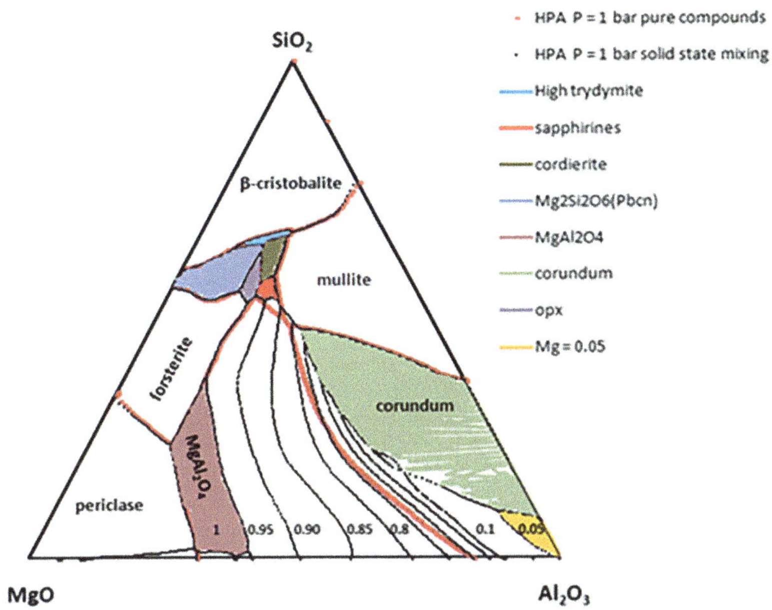


**Fig. 7.21** Primary phase boundaries of the MAS system at P = 1 bar. It is assumed that sapphirines mix ideally and that mullite is a pure compound. The numerals on the multiple oxide field are the MgO isopleths

The true eutectic of the system involves Mj-Pyr garnet, periclase, and corundum and occurs at T = 2803 K. The zoning of the Mj-Pyr mixture is similar to what was previously seen for the Opx(*Pbca*)—$AlAlO_3$ mixture, with progressively decreasing $Al_2O_3$ contents when moving toward the MgO-$SiO_2$ join. Superimposed on this phase diagram is the position of the KLB-1 peridotite investigated by Herzberg and Zhang (1996) and Trønnes and Frost (2002). According to these experiments, garnet is the first liquidus phase in KLB-1 peridotite and pyrolite bulk compositions at pressures of 22–23 GPa. We may tentatively simulate the process in the MAS analog. In doing this, we sum up MgO + FeO + MnO + CaO in the M vertex, $Al_2O_3$ + $Cr_2O_3$ in the A vertex, and $SiO_2$ + $TiO_2$ in the S vertex. According to our calculations, fractional stishovite crystallization of the MAS analog would bring the Mj-Pyr—stishovite univariant at T ~ 3250 K, then along the Mj-Pyr—periclase univariant to attain the eutectic finally. This confirms garnet as a predominant liquidus phase for ultrabasic bulk compositions at P–T conditions compatible with the mantle transition zone and the absence of ringwoodite, as experimentally observed.

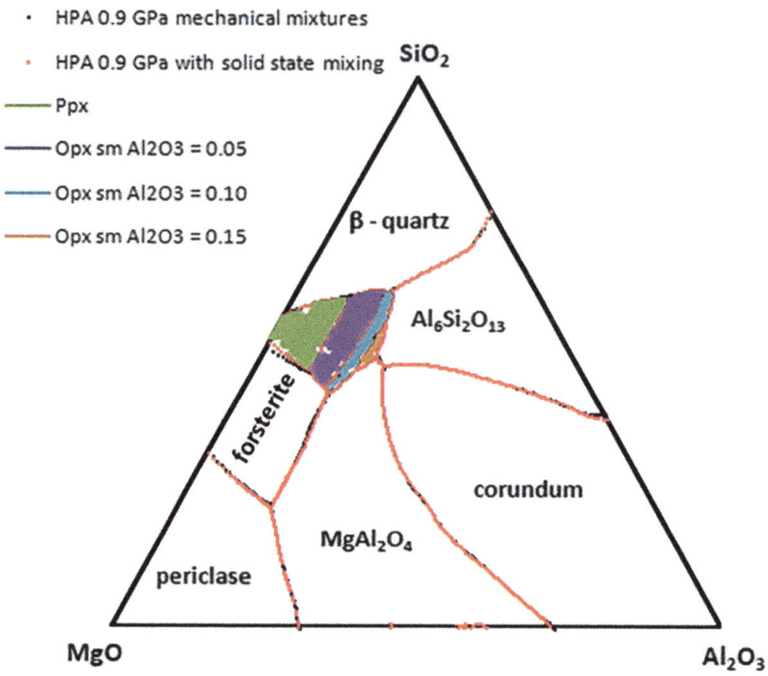


**Fig. 7.22** Primary phase boundaries of the MAS system at P = 0.9 GPa. Mullite solid-state mixing prevented

## 7.3 CaO-MgO-$SiO_2$ (CMS)

As in the case of CAS, and MAS also in the CaO-MgO-$SiO_2$ system, 3-components solids form in the inner part of the system. The new phases forming within CMS are diopside ($CaMgSi_2O_6$), akermanite ($Ca_2MgSi_2O_7$), merwinite ($Ca_3MgSi_2O_8$) and monticellite ($CaMgSiO_4$).

### *7.3.1 $CaMgSi_2O_6$ Pyroxenes*

Besides diopside, there are four fictive end-member polymorphs (necessary to depict the complex mixing properties with $Mg_2Si_2O_6$ end-members). The *C2c* pyroxenes *diopside* ($CaMgSi_2O_6$) has been carefully investigated by many Authors, due to its petrogenetic importance. The thermodynamic properties of the substance are, however, well defined only in the low-T range (calorimetric data up to ~1500 K). Although ab-initio calculations reproduce with high accuracy the vibrational properties of the substance (see Prencipe et al. 2012; and references therein) as far as we know, there has not been until now a serious attempt to perform a complete first principles characterization of the thermodynamic–thermophysical properties of this

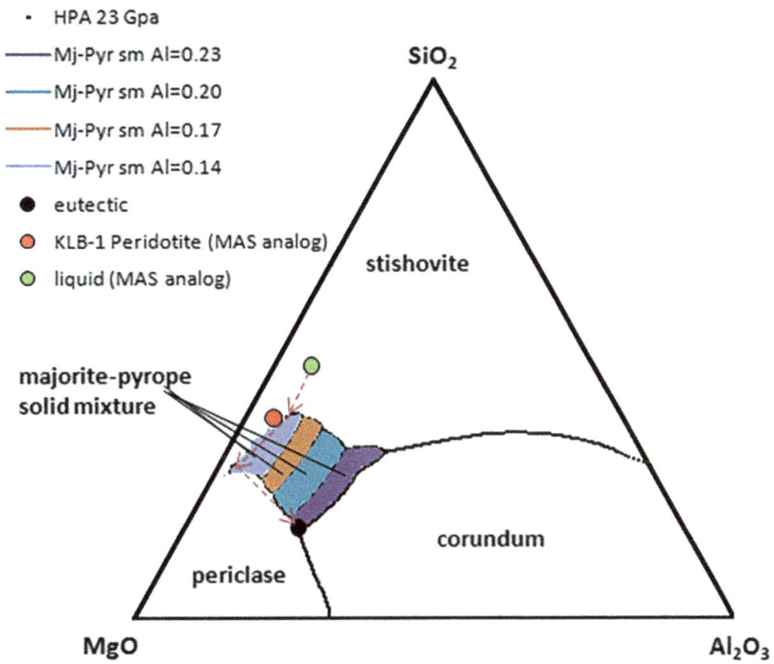


**Fig. 7.23** Primary phase boundaries of the MAS system at P = 23 GPa. The positions of the true eutectic of the system (T = 2841 K) and of the KLB-1 peridotite investigated by Herzberg and Zhang (1996) and Trønnes and Frost (2002) are superimposed. The red arrows mark schematically the crystallization path of the liquid along the tishovite primary phase field, then the stishovite—Mj-Pyr and Mj-Pyr—periclase univariants

holotropic substance. Therefore, we adopt here for $C_P = f(T)$ the parameterization of Berman (1988). For the thermophysical properties, we adopt the assessment of Saxena et al. (1993) with slight modifications on $K_0$, $K_1$, and $K_2$, necessary to achieve consistency with first principles at very high-T conditions (Fig. 7.24 and Table 4.2).

Table 7.3 resumes the thermodynamic properties of all $CaMgSi_2O_6$ polymorphs according to various assessments.

Our assessment is essentially based on the experimental observation of the Fo-Qtz-Di subsystem (see later) and on the assumption that the limited amounts of calcic components experimentally observed justify only a limited amount of conformational energy (purely entalpic).

We also assumed, for simplicity, that thermal expansion and compressibility are identical in the four polymorphs.

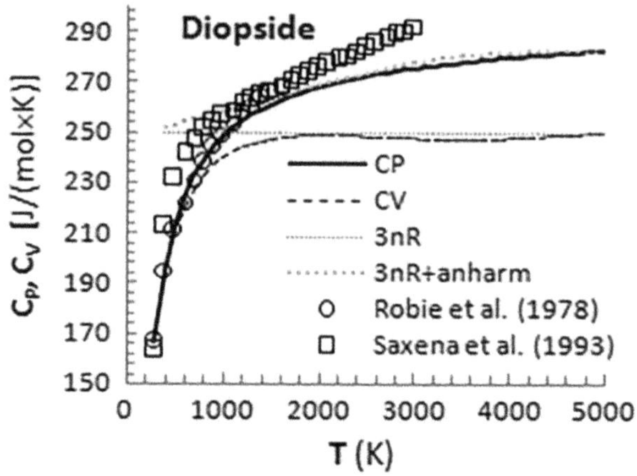


**Fig. 7.24** Isobaric heat capacity (heavy black solid line) and isochoric counterpart (black dashed line) of diopside. The $C_P$ assessments of Robie et al. (1978) and Saxena et al. (1993) are also plotted for comparative purposes

**Table 7.3** Standard-state properties of $CaMgSi_2O_6$ polymorphs according to various assessments. (1) Robie et al. (1978); (2) Berman (1988); (3) Saxema et al. (1993); (4) Shi et al. 1994); (5) Jung et al. (2005); (6) this work. $H°_{f,298}$ in kJ/mol; $S°_{298}$ in J/(mol × K); $V°_{298}$ in J/bar

| Space group | $H°_{f,298}$ | $S°_{298}$ | $V°_{298}$ | Refs. |
|---|---|---|---|---|
| *C2c* | −3210.760 | 143.09 | 6.609 | (1) |
| | −3200.583 | 142.50 | 6.620 | (2) |
| | −3201.820 | 143.10 | 6.610 | (3, 4) |
| | −3199.746 | 142.50 | 6.610 | (5) |
| | −3202.583 | 142.50 | 6.620 | (6) |
| *P2₁/c* | −3200.240 | 144.51 | 6.610 | (4) |
| | −3171.378 | 142.50 | – | (5) |
| | −3202.220 | 142.50 | 6.620 | (6) |
| *Pbca* | −3196.080 | 144.76 | 6.705 | (4) |
| | −3180.006 | 142.48 | – | (5) |
| | −3189.000 | 142.50 | 6.620 | (6) |
| *Pbcn* | −3191.250 | 144.94 | 6.562 | (4) |
| | −3182.675 | 142.50 | – | (5) |
| | −3185.512 | 142.50 | 6.620 | (6) |

## 7.3.2 $Ca_2MgSi_2O_7$ (Akermanite)

Akermanite ($Ca_2MgSi_2O_7$) is a tetragonal sorosilicate (space group $P4\ 2_1m$) belonging to the melilites series. It is a product of contact metamorphism of siliceous limestones and dolomites, first found pure at Monte Somma (Vesuvius). Akermanite undergoes a first-order transition at 358 K with a zero Clapeyron slope, zero $X_{\lambda 1}$ and $X_{\lambda 2}$, and $\Delta H_1{}^{st}{}_{order} = 452$ J/mol (Berman and Brown, 1985). We adopt here the

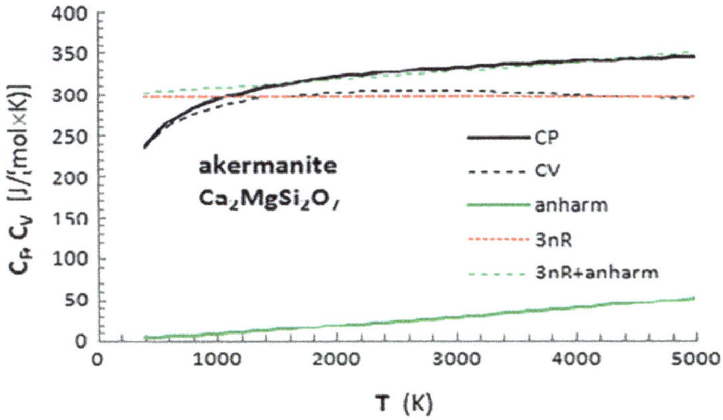


**Fig. 7.25** Isobaric heat capacity of akermanite (heavy black solid line) and isochoric counterpart (black dashed line)

thermodynamic data of Berman (1988) for $C_P = f(T)$, S°, and V° and lower its $H°_f$ of ~2.5 kJ/mol (Table 4.1). The adopted value ($H°_f = -3862.527$ kJ/mol) is halfway between the value tabulated by Berman (1988) (i.e., −3860.441 kJ/mol; apparently inconsistent with subsolidus equilibria; see later) and the calorimetric observations of Charlu et al. (1981) (−3865.6 ± 1.13 kJ/mol).

High-pressure synchrotron radiation powder and single-crystal diffraction investigations carried out by Merlini et al. (2009b) showed that the incommensurately modulated structure (IC of akermanite) transforms to a normal structure (N). The bulk modulus for the N structure, fitted with a Birch-Murnaghan EoS on powder data, gave $K_0 = 93.5(5)$ GPa, and is here adopted ($K' = 4$). We assume no thermal effects (i.e., $K_1$, $K_2 = 0$). For thermal expansion, we adopt the mean value experimentally observed by Merlini et al. (2005) by synchrotron X-ray powder diffraction (i.e., 32.1 × $10^{-6}$ $K^{-1}$).

With the adopted parameters, the isochoric and isobaric heat capacity of the substance obeys first principles (Fig. 7.25).

### 7.3.3 *$Ca_3MgSi_2O_8$ (Merwinite)*

Merwinite ($Ca_3MgSi_2O_8$) is a nesosilicate, crystallizing in the monoclinic system (space group $P2_1/b$) and present preferentially in HT-LP contact metamorphic zones with marbles. We adopt in this work the thermodynamic assessment of Berman (1988), which observed for this phase (and akermanite, monticellite as well), notable discrepancies in the heats of solution (Brousse et al. 1984; Neuvonen 1952a, b) and volume (Moore and Araki 1972; Robie et al. 1978). Thermal expansion is based on the tabulated values of Skinner (1966), and bulk modulus is based on the parameterization of Berman (1988) (see Eq. 6.4). The thermal effect on bulk modulus was obtained in such a way as to conform the computed anharmonicity to first principles (Fig. 7.26).

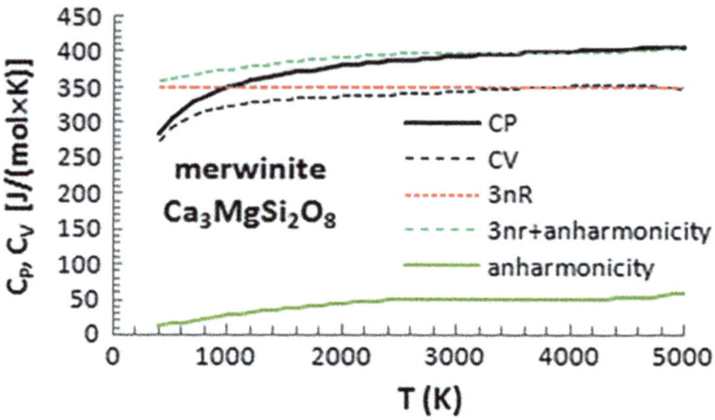


**Fig. 7.26** Isobaric heat capacity of merwinite (heavy black solid line) and isochoric counterpart (black dashed line)

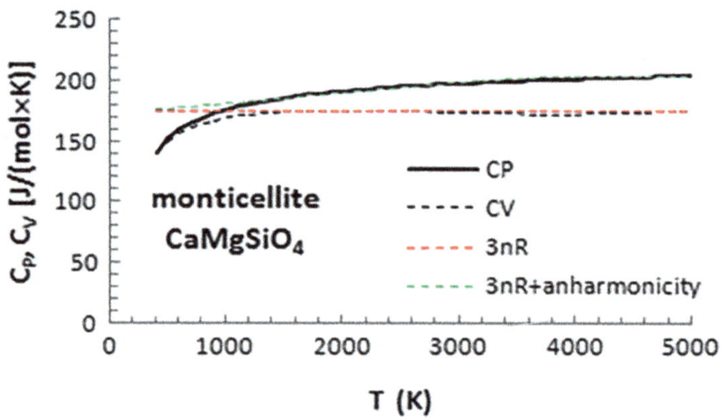


**Fig. 7.27** Isobaric heat capacity (heavy black solid line) and isochoric counterpart (black dashed line) of monticellite

## *7.3.4 $CaMgSiO_4$ (Monticellite)*

Monticellite is an orthorombic nesosilicate which may be conceived as a solid mixture of $Mg_2SiO_4$–$Ca_2SiO_4$ *Pbnm* end members. Just like dolomite, which is found as a stable compound of (Mg,Ca)$CO_3$ carbonate mixtures, the stability of $CaMgSiO_4$ is also due to the peculiar nature of mixing. With this provision, it is obvious that its thermodynamic-thermophysical parameters are constrained to a particularly stringent internal consistency with the end members.

For internal consistency, we adopt in this study the thermodynamic assessment of Berman (1988). Thermal expansion is based on the tabulated values of Skinner (1966), and bulk modulus is based on the parameterization of Berman (1988) (see Eq. 6.4). The thermal effect on bulk modulus was obtained in order to attain the Dulong-Petit limit in $C_V$ at high-T (Fig. 7.27).

#### 7.3.4.1 Subsolidus Equilibria

As done for the CAS and the MAS system, although in this study we are mainly interested in liquidus topology, we examined in some detail the phase equilibria among the ternary CMS phases at subsolidus as well (Fig. 7.28).

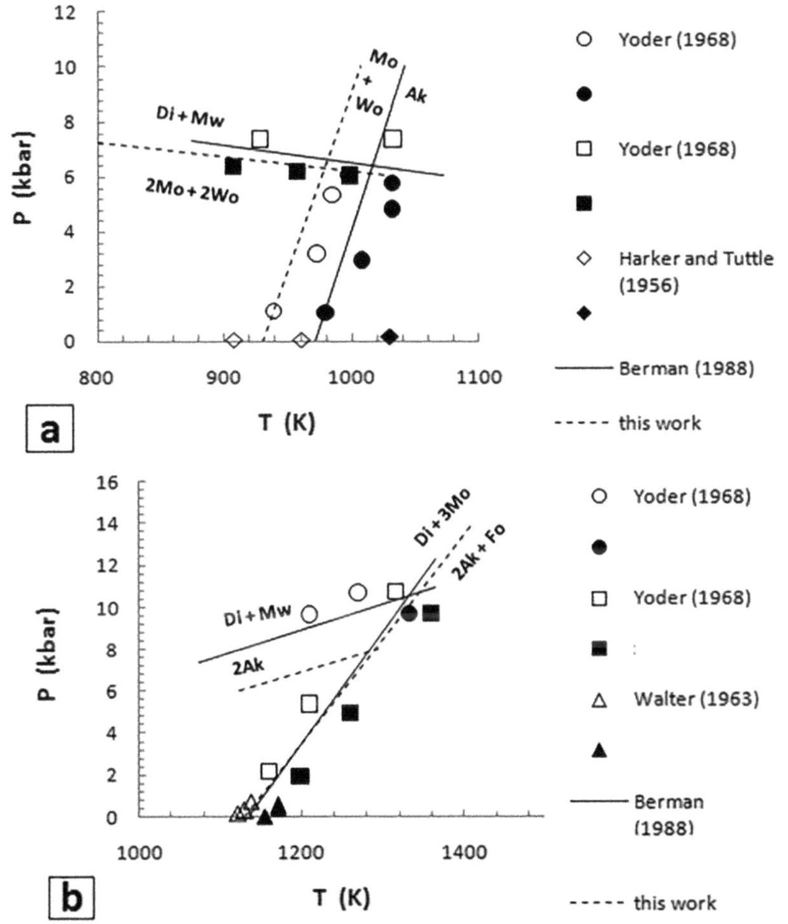


**Fig. 7.28** Computed phase equilibria with respect to experimental observations and the assessments of Berman (1988) in the CMS system. Open + filled symbols are reversals

We noted these discrepancies in the equilibria involving akermanite. Since these discrepancies are not systematic, it is difficult to establish where the problem resides (akermanite appears not stable enough in the Di + Mw ⇔ 2Ak equilibrium; too stable in the Mo + Wo ⇔ Ak equilibrium and appropriate in the Di + 3Mo ⇔ 2Ak + Fo equilibrium). These inconsistencies are, however, not quite marked and would not affect much the topology of the CMS liquidus.

#### 7.3.4.2 *CMS Liquidus: Topological Effects of Solid-State Mixing at P = 1 bar*

As anticipated, the optimization of the thermodynamic properties of pyroxenes required an accurate investigation of the Fo-Qtz-Di subsystem. This subsystem, due to its relevance in Earth Sciences, has been the object of accurate investigation by experimental petrologists for decades. In Fig. 7.29, we see the predictions of the HPA model compared with the assessments of Kushiro (1972) and Longhi and Budreau (1980). Our simulations were carried out with a discretization of 5 K on T and of 0.005 on solid mixtures, followed by 20 cycles of refinement. The dotted lines are the obtained univariant and invariant limits, and the colored fields return the identified primary phase at the liquidus. The four pyroxene polymorphs are identified sufficiently well for our purposes. We may particularly note that the diopside (*C2/c*) - pigeonite (Di/Pig) limit indicated by Longhi and Budreau (1980) is reproduced, but the polymorphic transition is more complex than expected, with a limit that is roughly

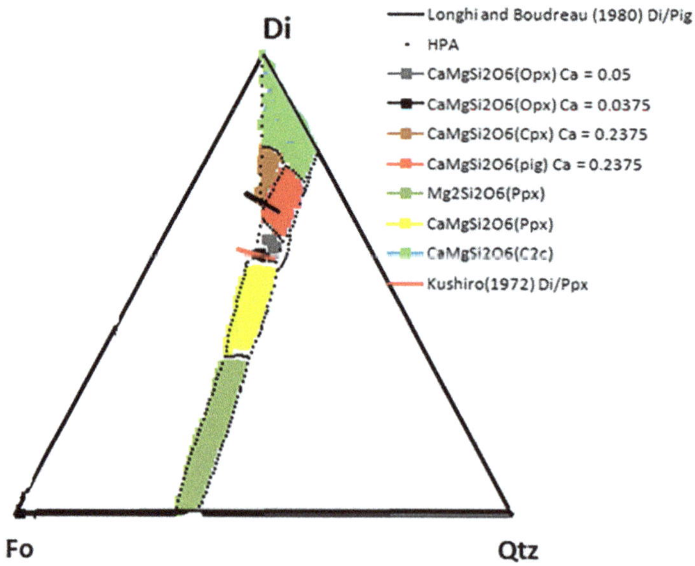


**Fig. 7.29** Primary phase boundaries of the Fo-Qtz-Di subsystem P = 1 bar. The HPA predictions are compared with the indications of Kushiro (1972) and Longhi and Budreau (1980)

parallel to the Fo-Di join rather than to the Fo-Qtz join as usually assumed. Moreover, the direct diopside (*C2/c*)—protopyroxene transition (Di/Ppx) indicated by Kushiro (1972) would rather be an orthopyroxene-protopyroxene (Opx/Ppx) transition.

It is quite evident that, by not taking into account the mixing behavior of the pyroxene polymorphs, one loses a lot of details on the actual nature of the aggregation state of the phase. Nevertheless, most part of the remaining primary phase fields at liquidus are essentially unvaried. We may now look at the entire CMS system at P = 1 bar (Fig. 7.30). The T range encompassed by the liquidus hypersurface at P = 1 bar is roughly 1600 K. The true eutectic of the system occurs at T = 1659 K, where high-tridymite–pseudo-wollastonite and diopside coexist with the liquid. The ternary is resolved simply with eight parameters. To the three parameters for chemical energy and four parameters for strain energy, it is simply added a subregular $A_{1,H}$ contribution of −30 kJ/mol for the CaO-MgO join (see Sect. 4.4.6). There cannot be in this system additional high-order interaction terms ascribable to polarization effects.

In Fig. 7.31, we see in more detail the primary phase fields obtained with Delaunay fragmentation and convex-hull analysis of the compositional space of interest.

In Table 7.4, the HPA ternary invariant points are compared with the results of the MQC assessment (Jung et al. 2004) and experimental observations.

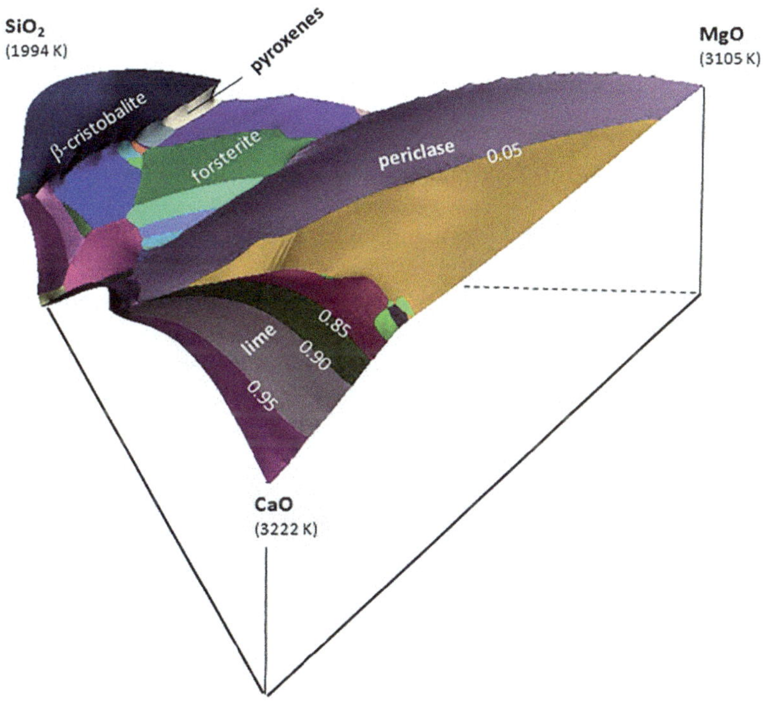


**Fig. 7.30** The liquidus surface of CMS at P = 1 bar. Step size = 5 K; compositional increments = 0.005. Smoothing runs = 20. Numerals in the “lime” and “periclase” primary phase fields are Ca isopleths

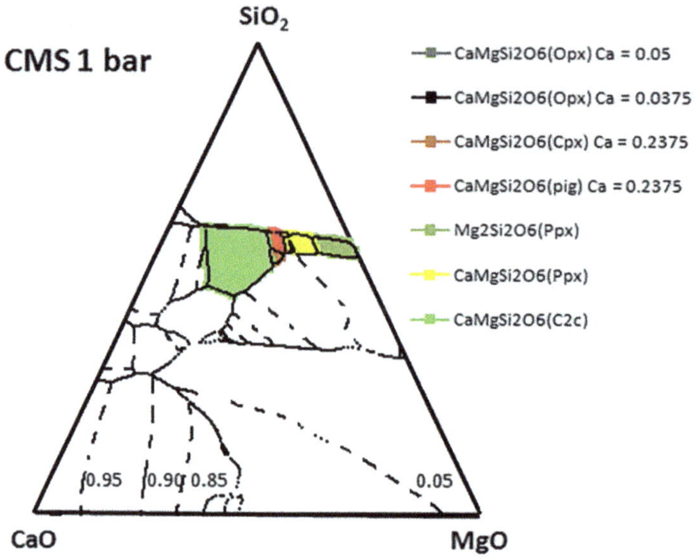


**Fig. 7.31** Primary phase boundaries of the CMS system at P = 1 bar. Numerals superimposed on the lime (left side) and periclase (right) primary phase fields are CaO isopleths

### 7.3.4.3 CMS at High P

The lack of experiments on the various $CaMgSi_2O_6$ polymorphs prevents an accurate evaluation of solid-state mixing properties at high pressure. We must thus limit ourselves to exploring the P–T topological effects induced on mechanical mixtures. In this respect, the petrological investigations carried out by various Authors on the Fo-$SiO_2$-Di subsystem are again of great help.

HPA computations carried out at P = 2, 5.1, and 7 GPa (Fig. 7.32) seem to indicate that solid-state mixing is not determinant in establishing the invariant loci.

The T predicted by HPA at invariance is almost within the experimental uncertainty, and the location of the invariant points is nearly coincident to what is observed experimentally. We may thus confidently assume that our CMS computations are informative, also neglecting solid-state miscibility.

Rising P to 5 GPa, important changes are observed: coesite replaces β-cristobalite; the fields of Ca-orthosilicates and merwinite expand; the field of Mg-orthosilicate shrinks (Fig. 7.33).

**Table 7.4** Calculated HPA (1) ternary invariants in the CMS system at P = 1 bar, versus experimental observations: (2) Yoder (1968); (3) Harfield et al. (1970); (4) Osborn (1942); (5) Longhi and Budreau (1980). The mean absolute bias on temperatures between HPA and experiments is 36.9 K

| Solid phases | | | Molar proportions | | | TT(K) | Refs. |
|---|---|---|---|---|---|---|---|
| A | b | c | CaO | MgO | SiO$_2$ | | |
| Periclase | Monticellite | Akermanite | 0.298 | 0.342 | 0.360 | 1775 | 2 |
| | | | 0.421 | 0.205 | 0.374 | 1789 | 1 |
| Merwinite | Monticellite | Akermanite | 0.374 | 0.244 | 0.382 | 1709 | 3 |
| | | | 0.494 | 0.135 | 0.372 | 1745 | 1 |
| | | | 0.489 | 0.140 | 0.371 | 1755 | 1 |
| Monticellite | Forsterite | Akermanite | 0.315 | 0.293 | 0.392 | 1703 | 3 |
| | | | 0.404 | 0.218 | 0.378 | 1779 | 1 |
| Akermanite | Diopside | Forsterite | 0.269 | 0.285 | 0.446 | 1630 | 3 |
| | | | 0.313 | 0.223 | 0.464 | 1697 | 1 |
| Akermanite | Diopside | Wollastonite | 0.351 | 0.175 | 0.474 | 1623 | 4 |
| | | | 0.373 | 0.144 | 0.483 | 1674 | 1 |
| Wollastonite (ps-woll) | Tridymite | Diopside | 0.308 | 0.112 | 0.580 | 1593 | 4 |
| | | | 0.316 | 0.074 | 0.610 | 1659 | 1 |
| Forsterite | Protopyroxene | Orthopyroxene | 0.117 | 0.358 | 0.525 | 1718 | 5 |
| | | | 0.144 | 0.296 | 0.559 | 1680 | 1 |
| Forsterite | Orthopyroxene | Pigeonite | 0.148 | 0.327 | 0.525 | 1683 | 5 |
| | | | 0.161 | 0.282 | 0.557 | 1670 | 1 |
| Forsterite | Diopside | Pigeonite | 0.170 | 0.310 | 0.520 | 1658 | 5 |
| | | | 0.166 | 0.280 | 0.553 | 1672 | 1 |
| Protopyroxene | Tridymite | Orthopyroxene | 0.127 | 0.310 | 0.563 | 1692 | 5 |
| | | | 0.133 | 0.273 | 0.594 | 1709 | 1 |
| Orthopyroxene | Tridymite | Pigeonite | 0.150 | 0.284 | 0.566 | 1660 | 5 |
| | | | 0.149 | 0.260 | 0.591 | 1674 | 1 |

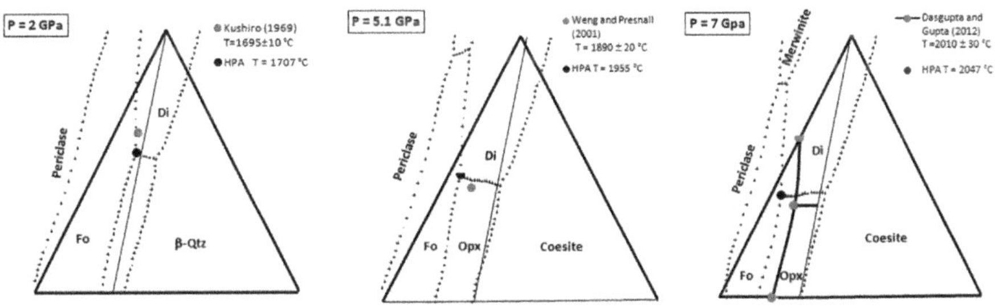


**Fig. 7.32** Primary phase boundaries of the Fo-Qtz-Di subsystem at pressure. The HPA predictions are compared with the indications of Kushiro (1969) at P = 2 GPa, Weng and Presnall (2001) at P = 5.1 GPa, and Dasgupta and Gupta (2012) at P = 7 GPa

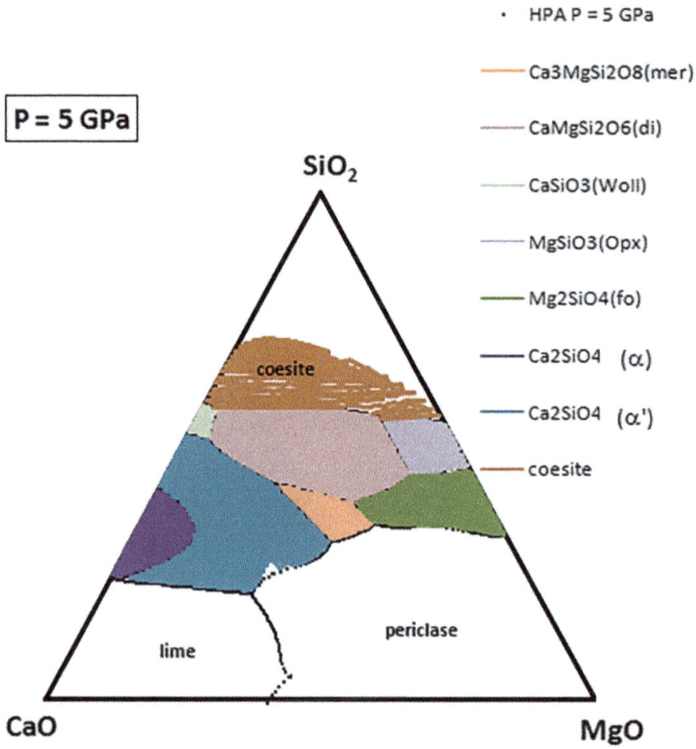


**Fig. 7.33** Primary phase boundaries of the CMS system at P = 5 GPa (mechanical mixtures)

We see finally in Fig. 7.34 the topological behavior of CMS at very high pressures. At P = 15 GPa, the stable silica polymorph is stishovite. A P = 25 GPa, a large part of the ternary liquidus s occupied by merwinite. Ca and Mg perovskites are also present. The primary phase fields of lime and periclase are not much affected by pressure.

## 7.4 CaO-MgO-$Al_2O_3$ (CMA)

The ternary compounds forming in this system are $Ca_3MgAl_4O_8$ (Welch 1961; Majumdar 1964), $CaMg_2Al_{16}O_{27}$, and $Ca_2Mg_2Al_{28}O_{46}$ (Göbbels et al. 1995; Iyi et al. 1995). Thermodynamic and thermophysical information on these compounds are scanty, and we have practically no hints about their eventual mixing behavior. So,

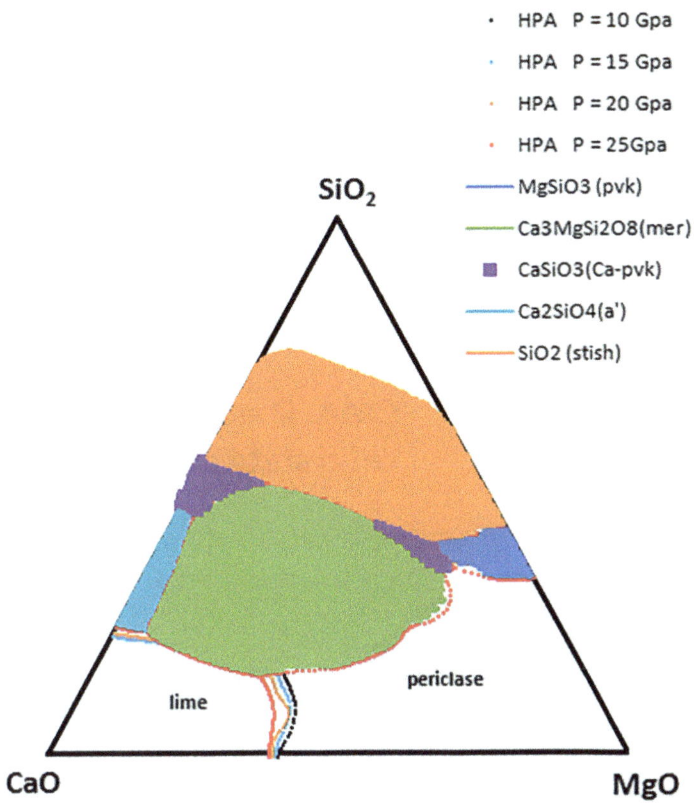


**Fig. 7.34** Primary phase boundaries of the CMS system at various pressures from 10 to 25 GPa (mechanical mixtures)

in the following, they will be treated as stoichiometric compounds. In their optimization of the CMA system, Jung et al. (2004) represented the isobaric heat capacity and the entropy of the substances as a simple summation of the properties of the constituent oxides. We adopt here their procedure and assume that volumes follow the same rules. Concerning the other thermophysical parameters we adopted $K_0$ = 200 Mbar $K' = 4$, $K_1 = -200$ bar/K, $K_3 = 0$, $K'' = 0$. Thermal expansion was obtained, as done for all the other solids of interest, imposing compliance with first principles (i.e., attainment of the isochoric heat capacity to the Dulong-Petit limit at high-T) (See Table 4.2).

### 7.4.1 CMA Liquidus at P = 1 Bar

Figure 7.35 is a 3D representation of the liquidus hypersurface of the CMA system at P = 1 bar conditions, based on a Delaunay fragmentation and convex-hull analysis of equilibrium loci with a T step of 5 K and an x step of 0.005, followed by 20 cycles of refinement. The ternary is resolved simply with ten parameters. To the four parameters for chemical energy and five parameters for strain energy, it is simply added a subregular $A_{1,H}$ contribution of −30 kJ/mol for the CaO-MgO join (see Sect. 4.4.6). As already anticipated for the CMS ternary, there cannot be in this system additional high-order interaction terms ascribable to polarization effects. The true

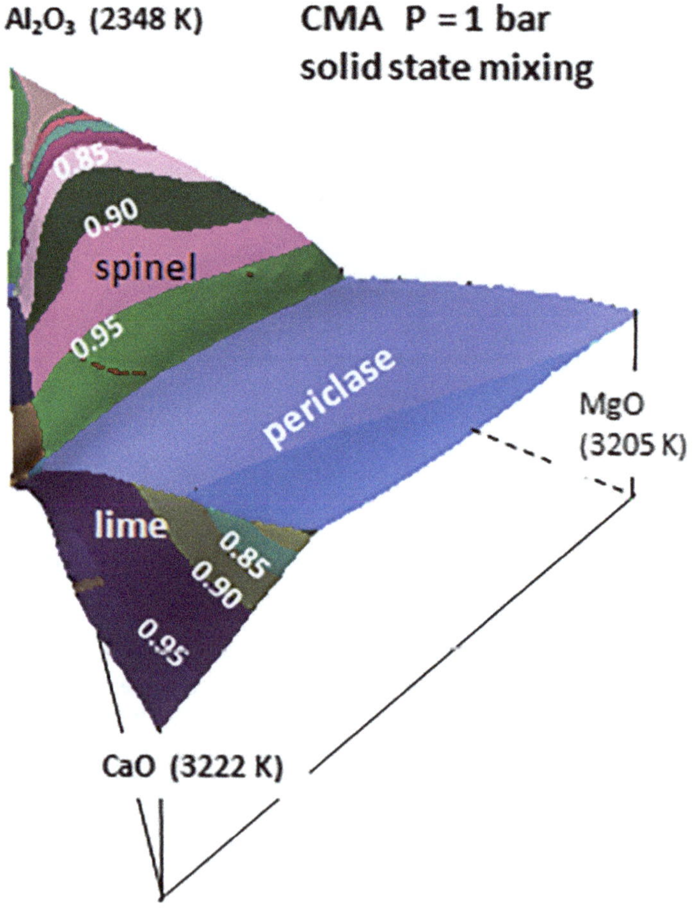


**Fig. 7.35** The liquidus surface of CMA at P = 1 bar. Step size = 5 K; compositional increments = 0.005. Smoothing runs = 20. Numerals in the "spinel" primary phase field are Mg isopleths; those on the 'lime' primary phase field are Ca isopleths

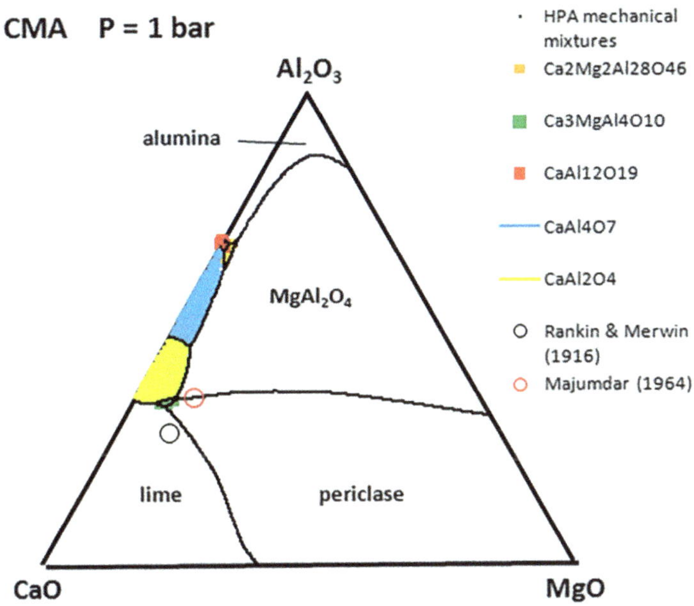


**Fig. 7.36** Primary phase boundaries of the CMA system at P = 1 bar. Step on T = 5 K, step on x = 0.005. The experimental observations of Rankin and Merwin (1916) and Majumdar (1964) are superimposed for comparative purposes

eutectic of the system is observed at T = 1710 K, where lime crystallizes together with $CaAl_2O_4$ and $Ca_3MgAl_4O_{10}$. The liquidus thus apparently spans a T-range of ~1400 K.

Figure 7.36 is the Schreinemaker's projection of equipotentials for stoichiometric phases. Accounting for solid-state mixing does not alter the univariant boundaries appreciably, nor the location of invariant points. It is of interest to note that only two invariants have been experimentally observed in the system until now: one observation is due to Rankin and Merwin (1916) (CaO-MgO-$Ca_3Al_2O_6$; T = 1723 K) and the other is due to Majumdar (1964) (MgO-$CaAl_2O_4$-$Ca_3MgAl_4O_{10}$; T = 1623 K).

The HPA invariant loci for pure phases are listed in Table 7.5. The energy imbalances with experiments and the MQC approach of Jung et al. (2004) are quite limited.

## *7.4.2 CMA at High P*

The lack of experiments at high P on various phases nucleating in this system prevents an accurate evaluation of solid-state mixing properties at high pressure. We must also anticipate that some thermophysical parameters are highly conjectural. HPA predictions at P = 1, 2, 5, 10, 15, 25 GPa when neglecting solid-state mixing are

**Table 7.5** Calculated HPA (1) ternary invariants in the CMA system at P = 1 bar, versus MQC calculations (2) (Jung et al. 2004) and experimental observations: (3) Rankin and Merwin (1916); (4) Majumdar (1964). Per = peritectic; eu = eutectic; ΔE = energy imbalance on the localized invariant point

| Solid phases | | | Type | T(K) | Moles | | | ΔE | Refs. |
|---|---|---|---|---|---|---|---|---|---|
| a | b | c | | | CaO | MgO | $Al_2O_3$ | | |
| $Ca_2Mg_2Al_{28}O_{46}$ | $CaMg_2Al_{16}O_{27}$ | $Al_2O_3$ ($CaAl_4O_7$) | per | 2029 | 0.292 | 0.020 | 0.688 | | 1 |
| | | | | 1981 | 0.272 | 0.066 | 0.662 | −1.871 | 2 |
| $Al_2O_3$($MgAl_2O_4$) | $CaAl_{12}O_{19}$ | $Ca_2Mg_2Al_{28}O_{46}$ | per | 2011 | 0.305 | 0.011 | 0.684 | | 1 |
| | | | | 2039 | 0.208 | 0.084 | 0.708 | 0.091 | 2 |
| $Ca_2Mg_2Al_{28}O_{46}$ | $CaAl_4O_7$ | $CaAl_{12}O_{19}$ | per | 1998 | 0.327 | 0.005 | 0.667 | | 1 |
| | | | | 1987 | 0.270 | 0.055 | 0.675 | −1.131 | 2 |
| $Ca_2Mg_2Al_{28}O_{46}$ | $MgAl_2O_4$ | $CaAl_4O_7$ | per | 1989 | 0.344 | 0.031 | 0.625 | | 1 |
| | | | | 1963 | 0.284 | 0.092 | 0.624 | −0.886 | 2 |
| $MgAl_2O_4$ | $CaAl_4O_7$ | $CaAl_2O_4$ | per | 1858 | 0.486 | 0.047 | 0.467 | | 1 |
| | | | | 1825 | 0.428 | 0.074 | 0.498 | 0.079 | 2 |
| $Ca_3MgAl_4O_{10}$ ($Ca_3Al_2O_6$) | CaO | MgO | per | 1769 | 0.597 | 0.068 | 0.336 | | 1 |
| | | | | 1723 | 0.618 | 0.103 | 0.279 | 0.329 | 3 |
| $MgAl_2O_4$ | MgO | $Ca_3MgAl_4O_{10}$ | per | 1748 | 0.566 | 0.081 | 0.354 | | 1 |
| | | | | 1622 | 0.536 | 0.111 | 0.353 | −1.634 | 2 |
| $MgAl_2O_4$ | $CaAl_2O_4$ | $Ca_3MgAl_4O_{10}$ | per | 1746 | 0.565 | 0.078 | 0.357 | | 1 |
| | | | | 1617 | 0.537 | 0.099 | 0.364 | −1.370 | 2 |
| CaO | $Ca_3MgAl_4O_{10}$ | $CaAl_2O_4$ | eu | 1710 | 0.614 | 0.043 | 0.344 | | 1 |
| ($Ca_3Al_2O_6$) | | | eu | 1582 | 0.606 | 0.049 | 0.345 | −0.222 | 2 |
| (MgO) | | | per | 1623 | 0.536 | 0.110 | 0.354 | −1.615 | 4 |

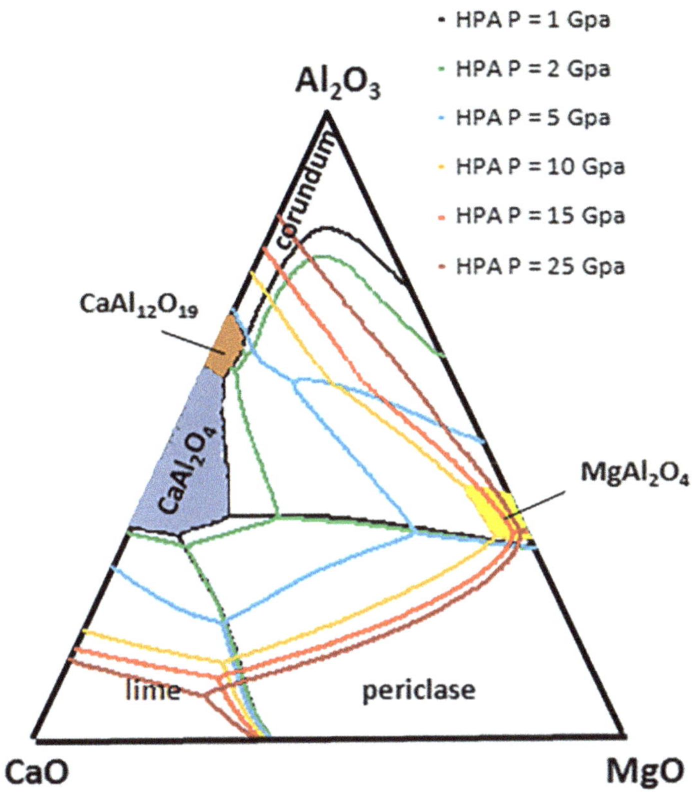


**Fig. 7.37** Primary phase boundaries of the CMA system at various pressures from 1 to 25 GPa (mechanical mixtures)

reproduced in Fig. 7.37. The first thing to note is that already at P = 1 GPa, ternary compounds are no longer present. Hibonite ($CaAl_{12}O_{19}$) is still present as a phase at liquidus. The primary phase field of $CaAl_2O_4$ progressively expands with pressure, predating the neighboring fields. At P = 5 GPa hibonite is no longer present while spinel ($MgAl_2O_4$) survives up to 10 GPa. From this pressure, only four phases are present at liquidus: besides the simplest compounds periclase, lime, and corundum, only $CaAl_2O_4$.

# Chapter 8
# 8 the $CaO$-$MgO$-$Al_2O_3$-$SiO_2$ Quaternary System

Having resolved the four limiting ternary faces of the system (second-rank simplexes), we may now form the tetrahedron (third-rank simplex). All forms of energy are self-propagating without distortion into the higher rank system, but the excess term defining the effect of the polarization work that depends upon localized partial charges. We anticipated that the simplest procedure is to assume that ionic polarizabilities and interionic distances do vary linearly with composition (see Eqs. 4.46–4.48). The computational approach is thus trivial. Another problem to resolve is the exponential growth of requested energy points with the ranking of the system. We found that a good compromise between accuracy and computational cost is a reduction of the T step from 5 to 10 K and an increase of the compositional step from 0.0005 to 0.001. The number of smoothing cycles is, in terms of computational effort, quite limited, and we kept 20 cycles as a good exercise.

## 8.1 Applications

Figure 8.1 is a screenshot of the CMAS tetrahedron generated in this way for $P = 1$ bar. In the upper part of the figure is shown the raw Delaunay + convex-hull conformation of univariant hypersurfaces in the system, and in the lower part represents the results of the smoothing cycles. The insets in the figure help to localize the discrete part of the third-rank simplex in terms of primary phase at liquidus. We may appreciate that no sensible distortion is introduced by the smoothing procedure (actually, this is not so for the $SiO_2$ apex, see later). The compositional discretization implies the creation of 970,000 × 200 energy points for the liquid and 58 × 200 energies for the solids (mechanical mixtures). The procedure identifies at this low pressure 197 invariant points (124 four components), emphasizing the need for mathematical derivations replacing simple (though smart) guesses based on a few experimental observations.

© The Author(s), under exclusive license to Springer Nature Switzerland AG 2026 
G. A. Ottonello et al., *Ab-initio Assisted Petrology*,
https://doi.org/10.1007/978-3-032-25658-4_8

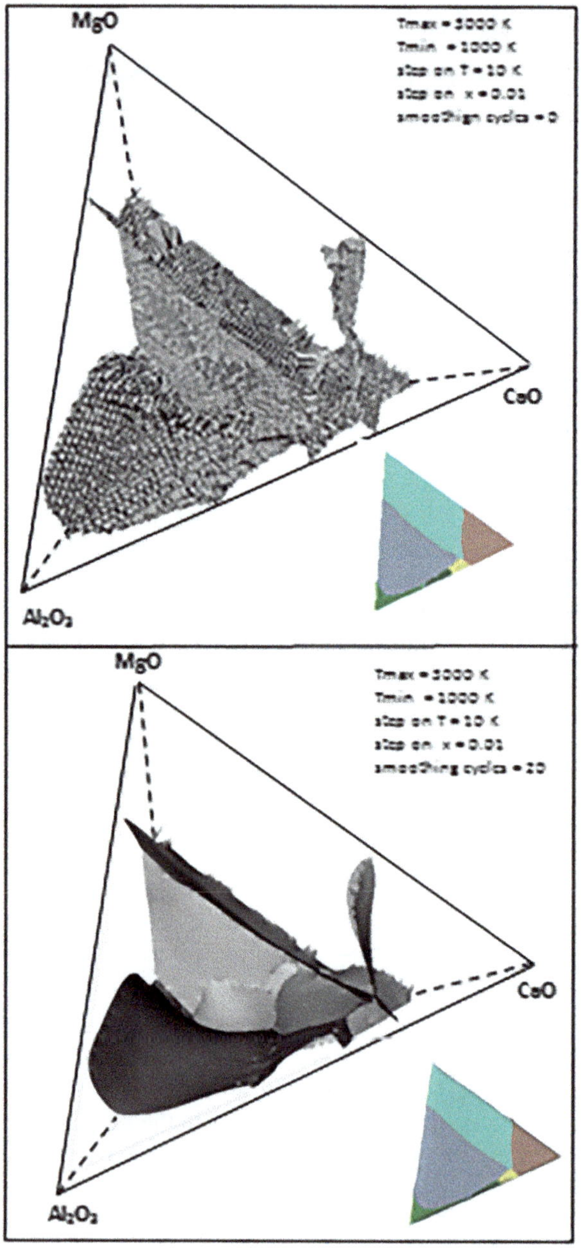


**Fig. 8.1** Screenshot of the CMAS system at P = 1 bar. HPA + Delaunay fragmentation and convex-hull analysis (NKCMAS.for + SIMPLICIAL.exe). In the upper part of the figure, the raw data after preliminary treatment, and in the lower part, the results after 20 smoothing cycles

In Fig. 8.2, we see how the four projections of the third-rank simplex compare with the discrete realizations of the second-rank simplexes at the same pressure. We may appreciate that the differences in T-steps (5 K in the ternaries and 10 K in the quaternary) and x-steps (0.005 in the ternaries and 0.01 in the quaternary) do not involve appreciable biases. We may note, however, that the smoothing introduces some perturbation in the neighborhood of the $SiO_2$ apex, ininfluent however in petrological implications. In the CMAS system at P = 1 bar, the Dalaunay fragmentation + convex-hull analysis generates 10,202 solid/liquid points of coexistence in the T range $1920 \leq T\ (K) \leq 3838$ ($1647 \leq T\ (°C) \leq 3535$).

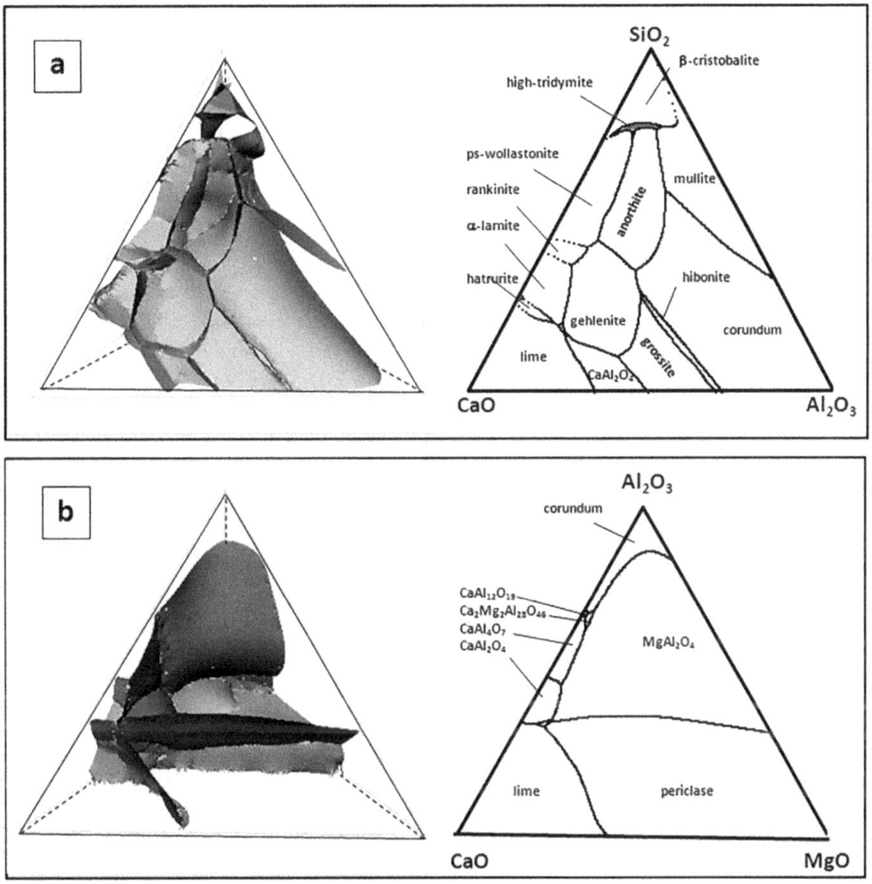


**Fig. 8.2** The four limiting faces of the CMAS system at P = 1 bar (left) and computed univariant boundaries and invariant points at the same pressure

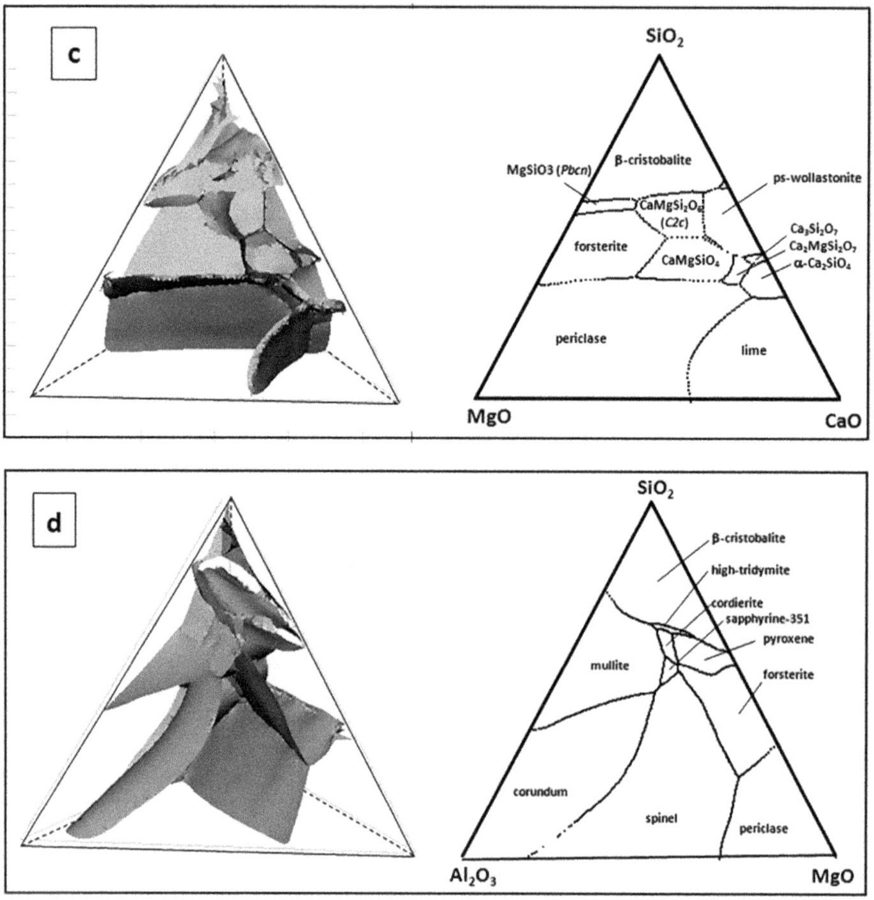


**Fig. 8.2** (continued)

## *8.1.1 Ccrystallization at Room Pressure*

In order to elucidate the petrological applications of the computed phase diagrams in practice, we present the case of fractional crystallization of a tholeiitic liquid (sample #9 of Presnall et al., 1978; see their Table 4.2). This composition is quite similar to the MORB medium value of Gale et al. (2013).

The first phase to form at liquidus is forsterite ($MgSi_2O_4$). The Delaunay fragmentation + convex-hull analysis generates 7157 points in the forsterite primary phase field (Fig. 8.3). Crystallization begins at $T = 1687$ K, and the residual liquid moves away from the initial composition along the direction identified by vectorial analysis of the barycentric coordinates (dashed line in Fig. 8.3).

To analyse the amount of heat loss necessary to fully induce crystallization, we would have at our disposal discrete liquid compositions placed exactly along

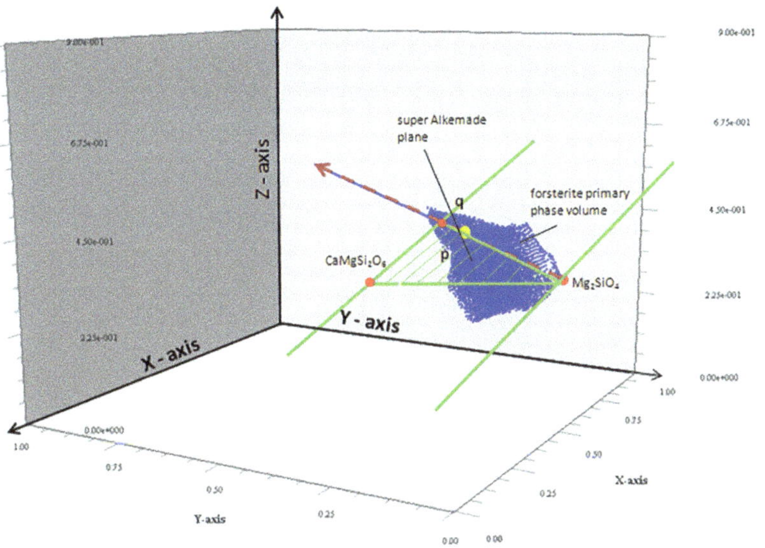


**Fig. 8.3** Initial steps of crystallization at P = 1 bar of a tholeiitic MORB. The residual liquid moves away from the initial composition along the vector until it attains the limit of the primary phase field (cotectic surface Forsterite—Anortite). The super-Alkemade plane, generated by the penetration point q through the divariant (Fo—Di) surface and the two crystallizing phases, allow to depict the binary path

the vectorial path, but this would be too computationally expensive, and we must simply exploit the thermodynamic information associated with the 7157 compositions retrieved by the system, selecting those ones nearer to the theoretical trajectory. A first analysis tells us that only 23% in moles of $MgSi_2O_4$ crystals may be subtracted before attaining the limits of the primary phase volume. The crystallization path may then be appropriately covered in 23 percentual steps, retrieving information from the nearby discrete liquid points at selected positions (initial—intermediate—final; Table 8.1). Once the residual liquid attains the divariant cotectic surface, a second phase ($CaAl_2Si_2O_8$, An) begins to coprecipitate with $Mg_2SiO_4$ (Fo). The path followed by the residual liquid along the divariant surface obeys a sort of "*super-Alkemade*" theorem. The original theorem of van Rijn van Alkemade is thus generalized for our third-rank simplex, accepting that the penetration point (**q** in Fig. 8.3) of the crystallization vector through the divariant surface is also a relative temperature maximum (see Fig. 8.1 in Eitel, 1951). This allows building a plane ("super-Alkemade plane" in Fig. 8.3) stemming from the three compositions at disposal (point **q**, primary phase crystals (Fo), and second-phase crystals (An); (see also Fig. 176 in Eitel, 1951).

The equation of the plane:

$$-0.06307X - 0.00732Y + 0.06441Z = 0.03574 \tag{8.1}$$

**Table 8.1** Post-processing of the HPA-generated database for P = 1 bar. The few steps along the crystallization path of a MORB-like CMAS analog (Presnall et al., 1978) depict the overall evolution from the primary phase field down to the invariant point. The T-lapse is quite restricted (80 K), but the amount of heat consumed (+) per unit-mole of magma increases progressively until the attainment of peritetic invariance, where reabsorption of Fo crystals causes a negative heat balance (-). The buoyancy is always positive for all crystals (which are thus expected to sink at the bottom of the reservoir). Temperatures and molar abundances in brackets are those actually found by the research algorithm (Postprocessor4#18.for), stemming from the smoothed coexistences identified by the SIMPLEX.exe package and operating on the data-files generated by the NKCMAS#21.for program with the NKCMAS_v26.in database. Thermodynamic data are retrieved by the post-processor code. Specific $H_{lat}$* is the heat of crystallization of the substance in the melt of the specified composition. Bulk $H_{lat}$* is the heat of crystallization of the solid paragenesis at invariance

| Phase | | Primary phase volume (Fo) | | | Cotectic surface (Fo-An) | | Univariant cotectic line (Fo-An-Di) | | Invariant (Fo-An-Di-En) |
|---|---|---|---|---|---|---|---|---|---|
| | T (K) | 1687 (1683) | 1654 (1653) | 1629 (1633) | 1628 (1623) | 1628 (1623) | 1624 (1623) | 1610 (1613) | 1602 (1603) |
| | % crystallization | 1.1 | 11.4 | 20.8 | 24.2 | 17.8 | 35.4 | 40 | – |
| Liquid | Moles | 0.989 | 0.886 | 0.792 | 0.758 | 0.822 | 0.646 | 0.611 | – |
| | CaO | 0.167(0.160) | 0.185 (0.180) | 0.212 (0.180) | 0.186 (0.190) | 0.193 (0.190) | 0.196 (0.200) | 0.189 (0.190) | 0.178 (0.180) |
| | MgO | 0.180 (0.180) | 0.129(0.130) | 0.050 (0.090) | 0.108 (0.110) | 0.107 (0.110) | 0.100 (0.100) | 0.073 (0.070) | 0.062 (0.060) |
| | $Al_2O_3$ | 0.091(0.090) | 0.101 (0.100) | 0.115 (0.110) | 0.120 (0.110) | 0.114 (0.110) | 0.112 (0.110) | 0.109 (0.110) | 0.107 (0.110) |
| | $SiO_2$ | 0.562 (0.570) | 0.586 (0.590) | 0.623 (0.620) | 0.585 (0.590) | 0.585 (0.590) | 0.593 (0.590) | 0.628 (0.630) | 0.652 (0.650) |
| | G (kJ/mol) | −1067.068 | −1083.586 | −1102.480 | −1091.878 | −1091.878 | −1093.263 | −1103.816 | −1107.554 |
| | H (kJ/mol) | −787.387 | −807.339 | −827.485 | −820.178 | −820.178 | −821.441 | −833.566 | −839.431 |
| | S [J/(mol × K)] | 166.165 | 167.103 | 168.383 | 167.391 | 167.391 | 167.466 | 167.529 | 167.248 |
| | V (J/bar) | 2.480 | 2.543 | 2.610 | 2.571 | 2.571 | 2.577 | 2.628 | 2.651 |
| | Cp [J/(mol × K)] | 84.389 | 85.388 | 86.718 | 86.255 | 86.255 | 86.164 | 86.674 | 86.961 |
| | Density (gr/cc) | 2.405 | 2.398 | 2.383 | 2.402 | 2.402 | 2.403 | 2.380 | 2.369 |
| Fo | Moles | 0.011 | 0.114 | 0.208 | 0.216 | 0.205 | 0.252 | 0.286 | −0.913 |
| | $H_{lat}$* (kJ/mol) | 141.333 | 159.394 | 178.283 | 170.349 | 170.349 | 171.612 | 183.110 | 188.216 |
| | V (J/bar) | 1.538 | 1.536 | 1.534 | 1.534 | 1.534 | 1.534 | 1.533 | 1.532 |

(continued)

**Table 8.1** (continued)

| Phase | | Primary phase volume (Fo) | | | Cotectic surface (Fo-An) | | Univariant cotectic line (Fo-An-Di) | | Invariant (Fo-An-Di-En) |
|---|---|---|---|---|---|---|---|---|---|
| | T (K) | 1687 (1683) | 1654 (1653) | 1629 (1633) | 1628 (1623) | 1628 (1623) | 1624 (1623) | 1610 (1613) | 1602 (1603) |
| | Density (gr/cc) | 3.050 | 3.054 | 3.057 | 3.058 | 3.058 | 3.058 | 3.060 | 3.061 |
| | Buoiancy (gr/cc) | 0.645 | 0.656 | 0.674 | 0.656 | 0.656 | 0.655 | 0.680 | 0.692 |
| | Specific $H_{\pm}$ (kJ) | 1.555 | 18.171 | 37.083 | 36.795 | 34.922 | 43.246 | 52.370 | −171.841 |
| An | Moles | 0 | 0 | 0 | 0.026 | −0.027 | 0.103 | 0.102 | 0.428 |
| | $H_{lat}$* (kJ/mol) | | | | −135.871 | −135.871 | −134.608 | −123.349 | −118.482 |
| | V (J/bar) | | | | 2.646 | 2.646 | 2.646 | 2.645 | 2.643 |
| | Density (gr/cc) | | | | 2.629 | 2.629 | 2.629 | 2.630 | 2.631 |
| | Buoiancy (gr/cc) | | | | 0.227 | 0.227 | 0.226 | 0.250 | 0.262 |
| | Specific $H_{\pm}$ (kJ) | | | | −3.533 | 3.669 | −13.865 | −12.582 | −50.710 |
| Di | Moles | 0 | 0 | 0 | | | −0.001 | 0.001 | 0.284 |
| | $H_{lat}$* (kJ/mol) | | | | | | 99.837 | 111.305 | 116.38 |
| | V (J/bar) | | | | | | 1.736 | 1.736 | 1.735 |
| | Density (gr/cc) | | | | | | 3.118 | 3.119 | 3.12 |
| | Buoiancy (gr/cc) | | | | | | 0.715 | 0.739 | 0.751 |

(continued)

**Table 8.1** (continued)

| Phase | | Primary phas∍ volume (Fo) | | | Cotectic surface (Fo-An) | | Univariant cotectic line (Fo-An-Di) | | Invariant (Fo-An-Di-En) |
|---|---|---|---|---|---|---|---|---|---|
| | T (K) | 1687 (1683) | 1654 (1653) | 1629 (1633) | 1628 (1623) | 1628 (1623) | 1624 (1623) | 1610 (1613) | 1602 (1603) |
| | Specific $H_\pm$ (kJ) | | | | | | −0.100 | 0.111 | 33.052 |
| En | Moles | 0 | 0 | 0 | | | | | 1.200 |
| | $H_{lat}$* (kJ/mol) | | | | | | | | 144.878 |
| | V (J/bar) | | | | | | | | 1.685 |
| | Density (gr/cc) | | | | | | | | 2.98 |
| | Buoiancy (gr/cc) | | | | | | | | 0.611 |
| | Specific $H_\pm$ (kJ) | | | | | | | | 173.854 |
| | Bulk $H_\pm$ (kJ) | 1.555 | 18.171 | 37.083 | 33.263 | 38.590 | 29.282 | 39.899 | −15.646 |

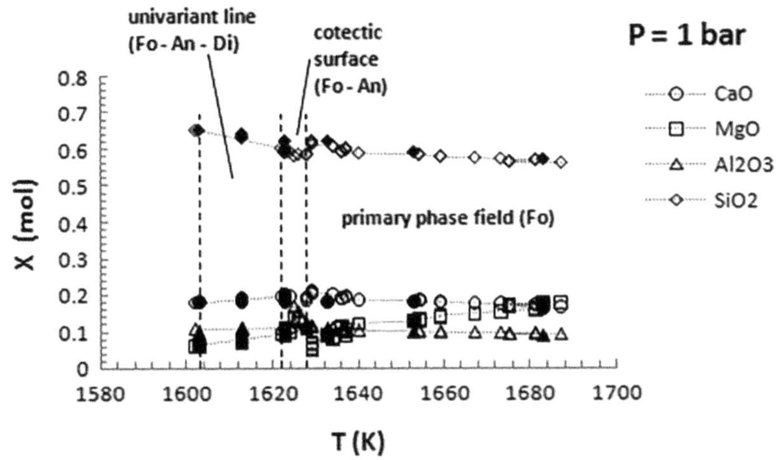


**Fig. 8.4** Chemical evolution of a MORB tholeiitic liquid (sample #9 of Presnall et al., 1978; see their Table 4.2). Open symbols are the explored compositions. Energy data generated by the NKCMAS#21.for code operating with the NKCMAS_v64.in database. Coexistences defined by the SIMPLEX.exe routine. Thermodynamic and thermophysical data at coexistence obtained with the POSTPROCESSOR4#18.for code. Filled symbols are the occurrences reported in Table 8.1

is obtained by Laplace expansion of the 3 × 3 determinant.

The super-Alkemade maximum is, by definition, also the first point on the Fo-An divariant cotectic. The analysis of the database shows that there are 3009 divariant points at disposal, but only 285 of them occur within the compositional limits of the super-Alkemade plane en that only few of them (58) obey with reasonable precision the cotectic path depicted by the theorem from the maximum down to the nearest univariant cotectic (Fo-An-Di) which is attained at $T = 1625$ K. The path along the univariant cotectic is then covered in six steps (two of them detailed in Table 8.1), and the invariant eutectic [Fo—An—Di(*C2/c*)—Opx(*Pbcn*)] is attained at 1603 K (1330 °C).

Quite significant is the additional information arising from the database: besides the molar amounts of precipitated (and, eventually, redissolved solids) obtainable by Gaussian inversion procedures, we retrieve other magnitudes of extreme importance in the petrological investigation of magmatic processes, i.e., heat content, entropy and molar volume of the coexisting phases, hence amount of heat consumed or released at each step, molar density and buoyancy (Table 8.1).

We may thus appreciate that the crystallization path is sufficiently continuous in terms of composition, temperature, and energy (Fig. 8.4).

Concerning the heat content, it is important to stress that the concept of "*latent heat*", introduced long ago by Joseph Black[1]: and of extreme importance in petrology, concerns the amount of enthalpy associated with a first-order transition among

[1] *Carey M (1807)* Black, Joseph. Lectures on the Elements of Chemistry: Delivered in the University of Edinburgh. *J. Robinson (Ed.), Pp.* 111–*112.*

isochemical aggregation states. As such, it implies a "*congruent*" process. Consider now, for instance, $Mg_2SiO_4$. At P = 1 bar, the crystalline polymorph (forsterite) melts congruently at 1893 K (see Par. 6.1.2). Our calculations restitute a "*latent heat of crystallization*" (i.e., $H^{\circ}_{f,crystal} - H^{\circ}_{f,liquid}$) of −52.911 kJ/mol on a 1-mol oxide basis (i.e., $Mg_{2/3}Si_{1/3}O_{4/3}$).

Only 84 K separates the beginning of crystallization within the Fo primary phase field from the invariant point. In this lapse of T, however, the amount of heat exchanged in crystallization increases progressively from ~ 1.6 to ~ 39.9 kJ per mole of magma until the attainment of the invariant point. Here, the massive reabsorption of forsterite crystals results in a negative heat balance (~ −15.6 kJ/mol), and the process may be expected to be facilitated by a sort of "*internal heating*" effect. At the invariant point, the heat exchanged in a closed system depends upon the heat transfer kinetics, and the amount of residual liquid is time-dependent. At all steps, all precipitated phases are denser than the liquid and may be expected to precipitate at the bottom of the reservoir. At the invariant point the molar composition [CaO = 0.18 (0.16); MgO = 0.06 (0.12); $Al_2O_3$ = 0.11 (0.15); $SiO_2$ = 0.65 (0.56)] is in overall agreement with the experimental observations of Presnall et al. (1978) (values in brackets) while the predicted T (1602 K) is higher than what experimentally observed (T = 1515 K; Presnall et al., 1978).

### *8.1.2 Crystallization at 2 GPa*

Basically, we aim to reproduce the experimental observations of Liu and Presnall (2000), which gave substantial input to mantle petrology with new hints on the genesis of andesites, the formation of eclogites and garnet clinopyroxenites, and the role of sapphyrine as a crucial petrogenetic marker. This exercise is quite demanding due to the complexity of the evolutionary path depicted by the Author through their experimental observations. Nothwithstanding this fact, here again we find a substantial agreement between experiments and calculus in terms of localization of divariant, univariant and invariant points though the accuracy is lower in the assessment of T. Based on their experiments Liu and Presnall (2000) delineated a complex trajectory of the residual liquid from the thernary (Fo—En—Di) invariant of Kushiro (1969) of the CMS system (T = 1968 ± 10 K) downward to the F invariant of the CMAS (a Fo—En—Di—Sp peritectic that the Authors locate at T = 1743 K (1470 °C). As shown by Liu and Presnall (2000) the reaction then further proceeds by heat loss to the invariant peritectic R [(En—Di—Sp—Sa; T = 1728 K); (1455 °C)], then the peritectic A [(En—Di _Saff—Gt; T = 1693 K); (1420 °C)] to attain finally the true eutectic T [(En—Di—An—Qz; T = 1623 K); (1350 °C)].

The ternary invariant (P in Fig. 4.3 of Kushiro, 1969; D in Fig. 8.3 of Liu and Presnall, 2000) is a peritectic reaction point in the CMAS system. The intrinsic nature of P is, according to Kushiro (1969), uncertain (eutectic or peritectic ?)

According to our calculations, point P is inside the primary phase field of forsterite (Fo), and a liquid of this composition begins to crystallize at 1997 K (1724 °C). The

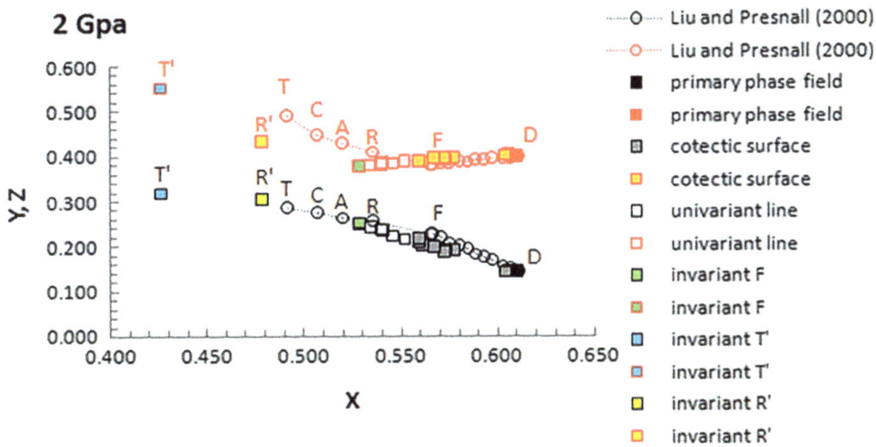


**Fig. 8.5** Chemical evolution at P = 2 GPa of a liquid of initial composition (D) equivalent to the MCS invariant according to Kushiro (1969). The results of HPA calculations (pure solids) are compared with the experimental observations of Liu and Presnall (2000). X, Y and Z are barycentric cohordinates: $X = MgO + 0.5{\cdot}(Al_2O_3 + SiO_2)$; $Y = MgO{\cdot}3^{-0.5}/2 + SiO_2{\cdot}3^{-0.5}/6$; $Z = SiO_2{\cdot}3^{-0.5}/6$

crystallization of Fo drives the residual liquid vectorially away from the starting composition, until the attainment of the Fo—Di cotectic (super-Alkemade point) at T = 1985 K (1712 °C). The univariant Fo—Di—En is then rapidly attained at T = 1935 K (1662 °C). The residual liquid moves then along this univariant line of descent and attains the peritectic invariant F (Di—En—Fo -Sp) at T = 1868 K (1595 °C). Invariant F in our calculations occurs in proximity to the invariant A of Liu and Presnall (2000) (see Fig. 8.5). As already seen for the previous investigated case (crystallization at P 1 bar), also in this case, the peritectic implies a substantial heat loss associated with reabsorption of Fo crystals (−132.199 kJ).

According to Liu and Presnall (2000), from the invariant F, the residual liquid would proceed along a complex trajectory which may involve the formation of sapphirine (Sa) or garnet (Gt) at intermediate stages (see the inset in their Fig. 3. 1). According to our calculations, sapphyrine (either 351 or 442 polytypes) never forms at this pressure, nor pyrope garnet, and the presence of an eventual garnet solid mixture is debatable due to positive enthalpic interactions in the mixture (see par. 4.4.8). Invariants R, A nd C cannot be thus reproduced. We find, however, a similar invariant R' with corundum replacing sapphyrine at T = 1813 K (1540 °C). Concerning the final invariant (i.e., the true eutectic T), again, we cannot reproduce the predicted solid assemblage (Di—En—An—Qz; T = 1350 °C) but a similar one, with corundum replacing β-Qz at T = 1752 K (1479 °C) (Fig. 8.5).

### 8.1.3 Toward the Mantle/core Boundary

Despite their importance in depicting early Earth's history (Canup, 2004), little is known of silicate liquids at pressures beyond those of the upper mantle. Published pre-heated shock wave experiments attained pressures of 40 GPa (Rigden et al., 1989), i.e., less than one third that of the mantle's base. Shock experiments have reached substantially higher pressures (Akins et al., 2004; Mosenfelder et al., 2007; Asimow and Ahrens, 2010), but their interpretation is affected by overdrive effects and solid–solid phase transitions along the Hugoniot.

Substantial improvements to the knowledge of the thermodynamic—thermophysical properties of silicate liquids in this little-explored realm have been provided by the FPMD investigations of Stixrude and Karki (2005), Stixrude et al. (2009) on $MgSiO_3$ liquids, and de Koker et al. (2013) on the MgO-$SiO_2$ system.

Because our thermodynamic—thermophysical assessment is parametrically quite complete and there are no implicit limits to the P–T realm that can be explored, we may tentatively consider the lowermost part of the mantle as well, as a final example of application of our computational approach.

Figure 8.6 compares the Mg-metasilicate liquidus in a wide P range obtained in this study and Ottonello (2024b) with the FPMD results of Stixrude and Karki (2005), Stixrude et al. (2009).

The comparison shows a substantial agreement between the two procedures. The proximity of the upper bound of the peridotite solidus at the mantle-core boundary (Holland and Ahrens, 1997) to the computed HPA eutectic pushed us to give a last potential model application to extreme conditions. The estimates of Holland and Ahrens (1997) agree well with the mantle solidus of Boehler (1993) (see also Par 7.3.2 in Ganguly, 2008). To achieve a sufficient accuracy in our estimates, we lowered the magnitude of the T steps to 5 K, duplicating the coexistence points and increasing the accuracy to 2 K. Figure 8.7 is a screenshot of the calculations at P = 136 GPa. At these extreme pressures, only six phases nucleate, namely periclase (MgO), corundum ($Al_2O_3$), lime (CaO), stishovite ($SiO_2$), perovskite ($MgSiO_3$), and ringwoodite ($Mg_2SiO_4$), but this last phase is never stable at liquidus for extrinsic stability reasons.

The eutectic is seen to occur at T = 4302 K (4130 °C) (CaO = 0.229, MgO = 0.007, $Al_2O_3$ = 0.758, $SiO_2$ = 0.006) and involves corundum (0.473 mol), lime (0.250 mol), periclase (0.027 mol), and perovskite (0.250 mol). The amount of heat consumed at melting is quite high (~2500 kJ/mol), and this implies that the lower mantle at the mantle/core limit would be substantially superheated to undergo partial fusion. Three phases under the current P,T conditions are potentially buoyant (corundum = −0.014 gr/cc; lime = −0.423 gr/cc; periclase = −0.082 gr/cc), and only perovskite is denser than the liquid (+0.476 gr/cc). The amount of liquid produced at the eutectic is, in any case, quite limited, and the liquid eventually moves along the lime-periclase-perovskite univariant, then the periclase-perovskite cotectic surface, to finally enter the periclase primary phase field, where the position of the Aluminum Depleted Komatiite (ADK) representative position is attained at T = 5862 K.

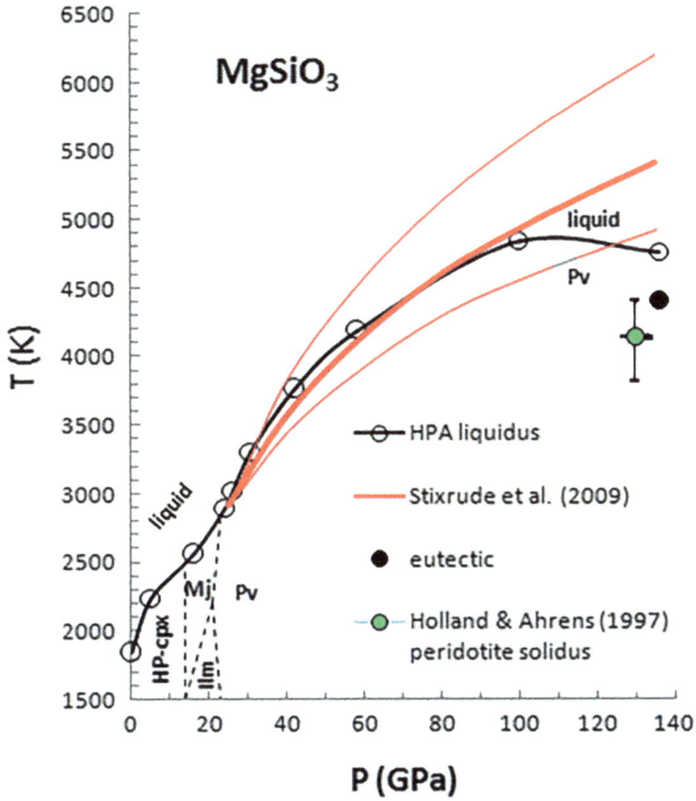


**Fig. 8.6** HPA $MgSiO_3$ liquidus (black empty circles and solid line) compared with the FPMD predictions of Stixrude et al. (2009) (red heavy solid line and limiting realm (thin red solid lines). The subsolidus HPA extrinsic limits of the various $MgSiO_3$ polymorphs. The black-filled circle is the eutectic of the system at P = 136 GPa. Pv = perovskite; Ilm = ilmenite; Mj = majorite. The upper bound of the peridotite solidus at the mantle-core boundary, according to Holland and Ahrens (1997), is also shown for comparative purposes

In situ X-ray diffraction measurements of $MgSiO_3$ carried out by Murakami et al. (2004) showed that Mg-perovskite has a polymorphic transition to a denser orthorhombic (*Cmcm*) structure at P,T conditions exceeding 125 GPa and 2500 K, i.e., approximately the zone of inflection of the HPA melting curve of the substance (cf. Figure 8.6). We do not, however, have enough thermodynamic-thermophysical data to include this polymorph in our provisional database, but we may anticipate that, most likely, its melting T would be somewhat higher than that of the corresponding *Pbnm* polymorph (Bridgmanite).

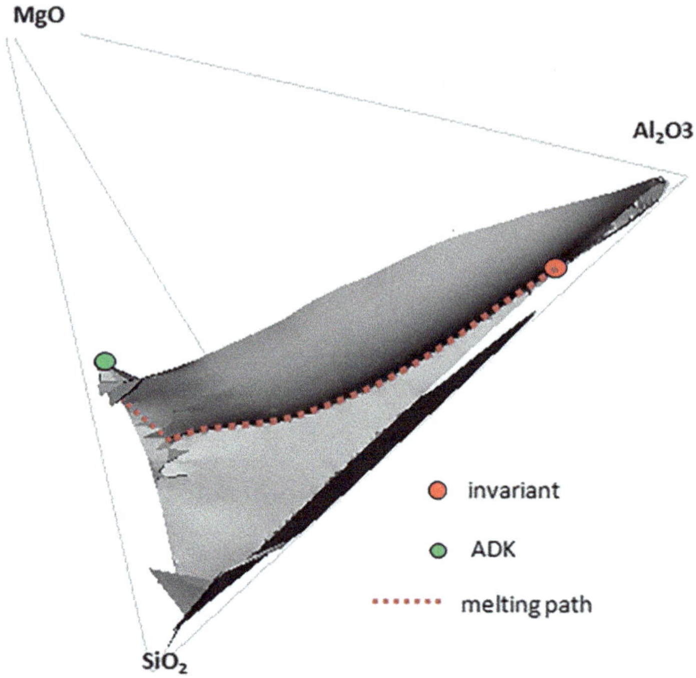


**Fig. 8.7** Snapshot of the CMAS system at P = 136 GPa. HPA + Delaunay fragmentation and convex-hull analysis (NKCMAS#21.for + SIMPLICIAL.exe). T step = 5 K; 20 smoothing cycles. The position of the eutectic invariant and of the selected final liquid (ADK) is highlighted. The dotted line marks the melting path

## 8.2 Concluding Remarks

The matter developed in this study belongs to a line of investigation that received a strong impulse from the pioneering studies of Connolly and Kerrick (1976), Connolly (1980). The "*piecewise-linear approximation*" adopted by the VERTEX program to build up subsolidus phase diagrams of petrogenetic relevance bears in fact close analogies with the "*Delaunay fragmentation*" adopted in this study.

The impact of Connolly and Kerrick's work on modern petrology has been humongous and beneficial, though most likely, not all potential users perceived the subtleties behind their apparently simple approach.

Let us hope that our work will be somewhat useful as well.

# References

Ackermann RJ, Sorrell CA (1974) Thermal expansion and the high-low transformation in quartz I High-temperature X-ray studies. J Appl Cryst 7:461–467
Adamkoviková K, Kosa L, Proks I (1980) The heat of fusion of $CaSiO_3$. Ceram Silik 24:193–201
Ahuja R, Belonoshko AB, Johansson B (1998) Melting and liquid structure of aluminum oxide using a molecular-dynamics simulation. Phys Rev E 57:1673–1676
Akaogi M, Yusa H, Shiraishi K, Suzuki T (1995) Thermodynamic properties of α-quartz, coesite, and stishovite and equilibrium phase relations at high pressures and high temperatures. J Geophys Res 100:22337–22347
Akaogi K, Hamada Y, Suzuki T, Kobayashi M, Okada M (1999) High pressure transitions in the system $MgAl_2O_4$-$CaAl_2O_4$: a new hexagonal aluminous phase with implication for the lower mantle. Phys Earth Planet Int 117:67–77
Akaogi M, Tanaka A, Ito E (2002) Garnet-ilmenite-perovskite transitions in the system $Mg_4Si_4O_{12}$-$Mg_3Al_2Si_3O_{12}$ at high pressures and high temperatures: phase equilibria, calorimetry and implications for mantle structure. Phys Earth Planet Int 132:303–324
Akaogi M, Yano M, Tejima Y, Iijima M, Kojitani H (2004) High-pressure transitions of diopside and wollastonite: phase equilibria and thermochemistry of $CaMgSi_2O_6$, $CaSiO_3$ and $CaSi_2O_5$-$CaTiSiO_5$ system. Phys Earth Planet Int 143:145–156
Akaogi M, Takayama H, Kojitani H, Kawaji H, Atake T (2007) Low-temperature heat capacities, entropies and enthalpies of $Mg_2SiO_4$ polymorphs, and α–β–γ and post-spinel phase relations at high pressure. Phys Chem Miner 34:169–183
Akins JA, Luo SN, Asimow PD, Ahrens TJ (2004) Shock-induced melting of $MgSiO_3$ perovskite and implications for melts in Earth's lowermost mantle. Geophys Res Lett 31:14
Alfé D (2005) Melting curve of MgO from first-principles simulations. Phys Rev Lett 94:235701
Alfè D (2009) PHON: a program to calculate phonons using the small displacement method. Comput Phys Commun 180:2622–2633
Alper AM, McNally RN, Ribbe PH, Doman RC (1962) The system MgO-$MgAl_2O_4$. J Am Ceram Soc 45:263–268
Anderson CT (1936) The heat capacities of quartz, cristobalite and tridymite at low temperatures. J Am Chem Soc 58:568–570
Anderson OL (1974) The determination of the volume dependence of the Gruneisen parameter. J Geophys Res 79:1153–1155
Anderson OL (1995) Equation of state for geophysics and ceramic sciences, oxford monographs on geology and geophysics, vol 31, Oxford University Press, 405p
Anderson OL, Isaak DG, Oda H (1991) Thermoelastic parameters of six minerals at high temperatures. J Geophys Res 96:18037–18046

© The Editor(s) (if applicable) and The Author(s), under exclusive license to Springer Nature Switzerland AG 2026

G. A. Ottonello et al., *Ab-initio Assisted Petrology*,
https://doi.org/10.1007/978-3-032-25658-4

Anderson OL, Zou K (1990) Thermodynamic functions and properties of MgO at high compression and high temperature. J Phys Chem Ref Data 19:69–81

Anderson OL, Isaak DG, Oda H (1991) Thermoelastic parameters of six minerals at high temperatures. J Geophys Res 96:18037–18046

Andrault D, Bolfan-Casanova N, Lo Nigro G, Bouhifd MA, Garbarino G, Mezouar M (2011) Solidus and liquidus profiles of chondritic mantle: implication for melting of the Earth across its history. Earth Planet Sci Lett 304:251–259

Angel RJ (1997) Transformation of fivefold-coordinated silicon to octahedral silicon in calcium silicate, $CaSi_2O_5$. Am Miner 82:836–834

Angel RJ, Hugh-Jones DA (1994) Equations of state and thermodynamic properties of enstatite pyroxenes. J Geophys Res 99(B10):19777–19783

Angel RJ, Ross NL, Seifert F, Fliervoet TF (1996) Structural characterization of pentacoordinate silicon in a calcium silicate. Nature 384:441–444

Angel RJ, Finger LW, Hazen RM, Kanzaki M, Weidner DJ, Liebermann RC, Veblen DR (1989) Structure and twinning of single-crystal $MgSiO_3$ garnet synthesized at 17 GPa and 1800 °C. Am Miner 74:509–512

Angel RJ, Kunz M, Miletich R, Woodland AB, Koch M, Knoche RL (1999) Effect of isovalent Si, Ti substitution on the bulk moduli of $Ca(T_{i1-}xS_ix)Si_O5$ titanites. Am Miner 84:282–287

Aramaki S, Roy R (1962) Revised phase diagram for the system $Al_2O_3$-$SiO_2$. J Amer Ceram Soc 5:229–242

Asimow PD, Ahrens TJ (2010) Shock compression of liquid silicates to 125 GPa: the anorthite-diopside join. J Geophys Res 115:B10209

Attene M (2004) Surface remeshing and applications. PhD Thesis DIST, Università di Genova

Aurenhammer F (1991) Voronoi diagrams—a survey of a fundamental geometric data structure. ACM Comput Surv 23:345–405

Austin JB (1954) The coefficient of linear thermal expansion of tridymite. J Am Chem Soc 76:6019–6020

Bajgain SK, Ghosh DB, Karki BB (2015) First-principles simulations of CaO and $CaSiO_3$liquids: structure, thermodynamics and diffusion. Phys Chem Miner 42:393–404

Barin I (1995) Thermochemical data of pure substances VCH, Weinheim, 1885p

Barnes P, Fentiman CH, Jeffery JW (1980) Structurally related dicalcium silicate phases. Acta Cryst A 26:353–356

Barron THK, Berg WT, Morrison JA (1959) On the heat capacity of crystalline magnesium oxide. Proc Roy Soc Lond 250:70–83

Bassett WA, Barnett JD (1970) Isothermal compression of stishovite and coesite up to 85 kilobars at room temperature by x-ray diffraction. Phys Earth Planet Inter 3:54–60

Bates FS, Fredrickson GH (1999) Block copolymers—designer soft materials. Phys Today 52:32–38

Belmonte D (2013) Ab initio thermodynamics of deep mantle minerals: the system MgO-$SiO_2$. PhD Thesis, University of Genova, Italy, 253p

Belmonte D (2017) First principles thermodynamics of minerals at HP-HT conditions: MgO as a prototypical material. Minerals 7:183–217

Belmonte D, Ottonello G, Vetuschi Zuccolini M (2013) Melting of α-$Al_2O_3$ and vitrification of the undercooled alumina liquid: ab-initio vibrational calculations and their thermodynamic implications. J Chem Phys 138:064507-1-64517

Belmonte D, Ottonello G, Vetuschi Zuccolini M (2014) Ab-initio thermodynamic and thermophysical properties of sapphirine end-members in the join $Mg_4Al_8Si_2O_{20}$–$Mg_3Al_{10}SiO_{20}$. Am Miner 99:1449–1461

Belmonte D, Gatti C, Ottonello G, Richet P, Vetuschi Zuccolini M (2016) Ab initio thermodynamic and thermophysical properties of sodium metasilicate, $Na_2SiO_3$, and their electron-density and electron-pair-density counterparts. J Phys Chem A 120:8881–8895

Belmonte D, Ottonello G, Vetuschi Zuccolini M (2017) Ab initio-assisted assessment of the CaO-$SiO_2$ system under pressure. Calphad 59:12–30

Belmonte D, Ottonello G, Vetuschi Zuccolini M, Attene M (2017) The system $MgO-Al_2O_3-SiO_2$ under pressure: a computational study of melting relations and phase diagrams. Chem Geol 461:54–64

Belonoshko AB, Dubrovinsky LS (1996) Molecular and lattice dynamics study of the $MgO-SiO_2$ system using a transferable interatomic potential. Geochim Cosmochim Acta 60:1645–1656

Belonoshko AB, Skorodumova NV, Rosengren A, Ahuja R, Johansson B, Burakovsky L, Preston DL (2005) High-pressure melting of $MgSiO_3$. Phys Rev Lett 94:195701

Belonoshko AB, Arapan S, Martonak R, Rosengren A (2010) MgO phase diagram from first principles in a wide pressure-temperature range. Phys Rev B 81:054110

Berman RG (1988) Internally-consistent thermodynamic data for minerals in the system $Na_2O-K_2O-CaO-MgO-FeO-Fe_2O_3-Al_2O_3-SiO_2-TiO_2-H_2O-CO_2$. J Petrol 29:445–522

Berman RG (1990) Mixing properties of Ca-Mg-Fe-Mn garnets. Am Miner 75:328–344

Berman RG, Brown TH (1984) A thermodynamic model for multicomponent melts, with application to the system $CaO-Al_2O_3-SiO_2$. Geochim Cosmochim Acta 48:661–678

Berman RG, Brown TH (1985) Heat capacity of minerals in the system $Na_2O-K_2O-CaO-MgO-FeO-Fe_2O_3-Al_2O_3-SiO_2-TiO_2-H_2O-CO_2$: representation, estimation, and high temperature extrapolation. Contrib Miner Petrol 89:168–183

Berman RG, Brown TH (1986) Erratum heat capacity of minerals in the system $Na_2O-K_2O-CaO-MgO-FeO-Fe_2O_3-Al_2O_3-SiO_2-TiO_2-H_2O-CO_2$: representation, estimation, and high temperature extrapolation. Contrib Miner Petrol 94:262

Berman RG, Brown TH, Greenwood H (1985) An internally consisten thermodynamic data base for minerals in the system $Na_2O-K_2O-CaO-MgO-FeO-Fe_2O_3-Al_2O_3-SiO_2-TiO_2-H_2O-CO_2$ Report N° TR-377. Atomic Energy of Canada Ltd, Ottawa

Blander M, Pelton AD (1983) Computer-assisted analyses of the thermodynamic properties of slags in coal-combustion systems. Argonne National Laboratory, Technical Report ANL/FE-83-19, Argonne, Illinois

Boates B, Bonev SA (2013) Demixing instability in dense molten $MgSiO_3$ and the phase diagram of MgO. Phys Rev Lett 110:135504

Boecke HE (1916) Grundlagen der physikalisch-chemischen Petrographie. Z Anorg Allg Chem 98:203–222

Boehler R (1982) Adiabats of quartz, coesite, olivine, and magnesium oxide to 50 kbar and 1000 K, and the adiabatic gradient in the Earth's mantle. J Geophys Res 87:5501–5506

Boehler R (1993) Temoeratures in the Earth's core from melting point measurements of iron at high pressures. Nature 363:534–536

Boehler R, Ramakrishnan J (1980) Experimental results on the pressure dependence of the Grüneisen parameter: a review. J Geophys Res 85:6996–7002

Boehler R, Chopelas A (1991) A new approach to laser heating in high pressure mineral physics. Geophys Res Lett 18:1147–1150

Boettcher AL (1970) The system $CaO-Al_2O_3-SiO_2-H_2O$ at high pressures and temperatures. J Petrol 11:337–379

Bohlen, SR, and Boettcher, L (1982) The quartz↔coesite transformation: a precise determination and the effects of other components J Geophys Res 87:7073–7078

Born M, Huang K (1954) Dynamical theory of crystal lattices. Oxford at the Claredon Press, 420p

Bose K, Ganguly J (1995) Quartz-coesite transition revisited: reversed experimental determination at 500–1200 °C and retrieved thermochemical properties. Am Miner 80:231–238

Bottinga Y, Richet P, Weill DF (1983) Calculation of the density and thermal expansion coefficient of silicate liquids. Bull Minér 106:129–138

Bourova E, Richet P (1998) Quartz and cristobalite: high-temperature cell parameters and volumes of fusion. Geophys Res Lett 25:2333–2336

Bourova E, Richet P, Petitet J-P (2006) Coesite ($SiO_2$) as an extreme case of superheated crystal: an X-ray diffraction study up to 1776 K. Chem Geol 229:57–63

Bowen NL, Andersen O (1914) The binary system $MgO-SiO_2$. Am J Sci 37:487–500

Bowen NL, Greig JW (1924) The system $Al_2O_3.2SiO_2$. J Am Ceram Soc 7:238–254

Brazhkin VV, McNeil LE, Grimsditch M, Bendeliani NA, Dyuzheva TI, Lityagina LM (2005) Elastic constants of stishovite up to its amorphization temperature. J Phys: Condens Matter 17:1869–1875

Brillouin L (1946) Wave propagation in periodic structures dover, New York, 274p

Brodholt J, Patel A, Refson K (1996) An Ab Initio study of the compressional behavior of forsterite. Am Miner 76:1100–1109

Brousse C, Newton RC, Kleppa OJ (1984) Enthalpy of formation of forsterite enstatite akermanite monticellite and merwinite at 1073 K determined by alkali borate solution calorimetry. Geochim Cosmochim Acta 48:1081–1088

Brown JM, Shankland TJ (1981) Thermodynamic parameters in the earth as determined from seismic profiles. Geophys J Int 66:579–596

Bruno E, Chiari G, Facchinelli A (1976) Anorthite quenched from 1530 °C I Structure refinement. Acta Cryst B32:3270–3280

Burt JB, Ross NL, Angel RJ, Koch M (2006) Equations of state and structures of andalusite to 98 GPa and sillimanite to 85 GPa. Am Miner 91:319–326

Cámara F, Bindi L, Pagano A, Pagano R, Gain SEM, Griffin WL (2019) Dellagiustaite: a novel natural spinel containing $V^{2+}$. Minerals 9:4–20

Cameron M, Papike JJ (1981) Structural and chemical variations in pyroxenes. Am Miner 6:1–50

Canup RM (2004) Simulations of a late lunar-forming impact. Icarus 168:433–456

Carpenter MA, Ferry JM (1984) Constraints on the thermodynamic mixing properties of plagioclase feldspars. Contrib Miner Petrol 87:138–148

Carpenter MA, McConnell JD (1984) Experimental delineation of the transformation in intermediate plagioclase feldspars. Am Miner 69:112–121

Cartz L, Jorgensen JD (1981) The high-pressure behavior of α-quartz, oxynitride, and nitride structures. J Appl Phys 52:236–244

Castanet R (1984) Selected data on the thermodynamic properties of α-alumina. High Temp High Press 16:449–457

Charlu TV, Newton RC, Kleppa OJ (1975) Enthalpies of formation at 970 K of compounds in the system $MgO$-$Al_2O_3$-$SiO_2$ by high temperature solution calorimetry. Geochim Cosmochim Acta 39:1487–1497

Charlu TV, Newton RC, Kleppa OJ (1978) Enthalpy of formation of some lime silicates by high-temperature solution calorimetry, with discussion of high pressure phase equilibria. Geochim Cosmochim Acta 42:367–375

Charlu TV, Newton RC, Kleppa OJ (1981) Thermochemistry of synthetic $Ca_2Al_2SiO_7$ (gehlenite)—$Ca_2MgSi_2O_7$ (akermanite) melilites. Geochim Cosmochim Acta 45:1609–1617

Chase MW Jr (1998) NIST-JANAF thermochemical tables journal of physics and chemistry reference data monograph no 9 American chemical society. American Institute of Physics and National Institute of Standards and Technology, 1951p

Chatterjee ND (1989) An internally consistent thermodynamic data base on minerals: applications to the Earth's crust and upper mantle. PhD thesis, City University of New York

Chatterjee ND, Johannes W, Leistner H (1984) The system $CaO$-$Al_2O_3$-$SiO_2$-$H_2O$: new phase equilibria data, some calculated phase relations, and their petrological applications. Contrib Miner Petrol 88:1–13

Chaudhuri SP (1987) Melting/decomposition of mullite: incongruent or congruent? I phase equilibria of the system $Al_2O_3$-$SiO_2$. Ceram Int 13:167–175

Chen C-H, Presnall DC (1975) The system $Mg_2SiO_4$-$SiO_2$ at pressures up to 25 kilobars. Am Miner 60:398–406

Chernosky JV Jr, Day HW, Caruso LJ (1985) Equilibria in the system $MgO$-$SiO_2$-$H_2O$: I experimental determination of the stability of Mg-anthophyllite. Am Miner 70:223–236

Chiari G, Facchinelli A, Bruno E (1978) Anorthite quenched from 1630 °C discussion. Acta Cryst B 34:1757–1764

Chopelas A, Boehler R, Ko T (1994) Thermodynamics and behavior of $\gamma$-$Mg_2SiO_4$ at high pressure: implications for $Mg_2SiO_4$ phase equilibrium. Phys Chem Miner 21:351–359

Christy AG, Phillips BL, Güttler BK, Kirkpatrick RJ (1992) A $^{27}$Al and $^{29}$Si MAS NMR and infrared spectroscopic study of Al-Si ordering in natural and synthetic sapphirine. Am Miner 77:8–18

Chibisov AN (2013) Effect of pressure on the atomic and electronic structure of enstatite $MgSiO_3$: Ab initio calculations. Zh Tekh Fiz 83:69–72

Cohen LH (1979) High-low cristobalite transitions in $SiO_2$, $AlPO_4$ and $GaPO_4$ Investigations by differential thermal analysis under hydrostatic pressure <6 kbar. Phil Mag 39A:399–404

Cohen LH (1980) Tridymite: effect of hydrostatic pressure to 6 kbars on temperatures of two rapidly reversible transitions. Contrib Miner Petrol 71:401–405

Cohen LH, Klement W Jr (1967) High-low quartz inversion: determination to 35 kilobars. J Geophys Res 72:4245–4251

Cohen LH, Klement W Jr (1980) Tridymite: effect of hydrostatic pressure to 6 kbars on temperatures of two rapidly reversible transitions. Contrib Miner Petrol 71:401–405

Cohen RE, Weitz JS (1998) The melting curve and premelting of MgO In: Manghnani MH, Yagi T (eds) Properties of earth and planetary materials at high pressure and temperature. American Geophysical Union Washington, pp 185–196

Connolly JAD (1990) Multivariable phase diagrams: an algorithm based on generalized thermodynamics. Am J Sci 290:666–718

Connolly JAD, Kerrick DM (1987) An algorithm ad computer program for calculating composition diagrams. Calphad 11:1–54

Coughlin JP (1956) Heat of formation of crystalline $CaO{\cdot}Al_2O_3$, $12CaO{\cdot}7Al_2O_3$, $3CaO{\cdot}Al_2O_3$. J Amer Chem Soc 78:5479–5482

Coughlin JP, O'Brien CJ (1957) High temperature heat contents of calcium orthosilicate. J Phys Chem 61:767–769

Dalton JA, Presnall DC (1997) No liquid immiscibility in the system $MgSiO_3$-$SiO_2$ at 50 GPa. Geochim Cosmochim Acta 61:2367–2373

D'Amour H, Denner W, Schultz H (1979) Structure determination of α-quartz up to $68\times10^8$ Pa. Acta Crist B35:550–555

Danckwerth A, Newton RC (1978) Experimental determination of the spinel peridotite to garnet peridotite reaction in the system MgO-$Al_2O_3$-$SiO_2$ in the range 900°–1100°C and $Al_2O_3$ isopleths of enstatite in the spinel field. Contrib Miner Petrol 66:189–201

Dasgupta S, Gupta AK (2012) The forsterite-diopside-silica systema at 7 GPa and its significance in the melting behavior of the mantle. Can Min 50:1243–1253

Da Silva C, Stixrude L, Wentzcovitch RM (1997) Elastic anisotropy of forsterite at high pressure. Geophys Res Lett 24:1963–1966

Davidson PM, Lindsley DH (1989) Thermodynamic analysis of pyroxene-olivine-quartz equilibria in the system MgO-FeO-$SiO_2$. Am Miner 74:18–30

Deer WA, Howie RA, Zussman J (1997) Rock-forming minerals, single-chain silicates, vol 2A, 2nd edn. The Geological Society, London, p 668p

de Koker N, Stixrude L (2009) Self-consistent thermodynamic description of silicate liquids, with application to shock melting of MgO periclase and $MgSiO_3$ perovskite. Geophys J Int 178:162–179

de Koker N, Karki BB, Stixrude L (2013) Thermodynamics of the MgO-$SiO_2$ liquid system in Earth's lowermost mantle from first principles. Earth Planet Sc Lett 361:58–63

De La Pierre M, Belmonte D (2016) Ab initio investigation of majorite and pyrope garnets: lattice dynamics and vibrational spectra. Am Miner 101:162–174

Demuth T, Jeanvoine Y, Hafner J, Angyan JG (1999) Polymorphism in silica studied in the local density and generalized-gradient approximations. J Phys Condens Matter 11:3833–3874

Dmitriev V, Torgashev V, Tolédano P, Salje EKH (1997) Theory of $SiO_2$ polymorphs. Europhys Lett 37:553–558

Ditmars DA, Douglas TB (1971) Measurement of the relative enthalpy of pure α-$Al_2O_3$ (NBS heat capacity and enthalpy. Standard reference material No 720) from 273 to 1173 K. J Res Natl Bur Stand 75A:401–420

Ditmars DA, Ishihara S, Chang SS, Bernstein G, West ED (1982) Enthalpy and heat-capacity standard reference material: synthetic sapphire (Alpha-$Al_2O_3$) From 10 to 2250 K. J Res Natl Bur St 87:159–163

Doman RC, Barr JB, McNally RN, Alper AM (1963) Phase equilibria in the system CaO-MgO. J Am Ceram Soc 46:313–316

Dovesi R, Saunders VR, Roetti C, Orlando R, Zicovich-Wilson CM, Pascale F, Civalleri B, Doll K, Harrison NM, Bush IJ, D'Arco P, Llunell M (2009) CRYSTAL09 user's manual. Università di Torino, Torino, Italy

Downs RT, Palmer DC (1994) The pressure behavior of α cristobalite. Am Miner 79:9–14

Dovesi R, Saunders VR, Roetti C, Orlando R, Zicovich-Wilson CM, Pascale F, Civalleri B, Doll K, Harrison NM, Bush IJ, D'Arco Ph, Llunell M, Causà M, Noël Y (2014) CRYSTAL14 User's Manual Università di Torino, Torino, 382p

Drewitt JWE, Barnes AC, Jahn S, Kohn SC, Walter MJ, Novikov AN, Neuville DR, Fischer HE, Hennet L (2017) Structure of liquid tricalcium aluminate. Phys Rev B 95:064203

Du Z, Lee KKM (2014) High-pressure melting of MgO from (Mg,Fe)O solid solutions. Geophys Res Lett 41. https://doi.org/10.1002/2014GL061954

Duan W, Karki B, Wentzcovitch RM, Gu BL (1991) Ab initio study of $MgSiO_3$ low-clinoenstatite at high pressure. Am Miner 86:762–766

Dubrovinsky LS, Saxena SK (1997) Thermal expansion of periclase (MgO) and tungsten (W) to melting temperatures. Phys Chem Miner 24:547–550

Eitel W (1951) Silicate melt equilibria Rutgers University Press, New Jersey, 159p

Elliott JF, Gleiser M (1960) Thermochemistry for steelmaking. Addison-Wesley, Boston, p 296p

Elsner von GronowH, Schwiete HE (1933) Die spezifischenWärmen fon CaO, $Al_2O_3$, CaO-$Al_2O_3$, 3CaO-$Al_2O_3$, 2CaO-$SiO_2$, 3CaO-$SiO_2$, 2CaO-$Al_2O_3$-$SiO_2$ von 20 bis 1500 °C. Z Anorg Allg Chem 216:185–195

Erba A, Mahmoud A, Belmonte D, Dovesi R (2014) High pressure elastic properties of minerals form Ab initio simulations: the case of pyrope, grossular and andradite silicate garnets. J Chem Phys 140:124703

Erba A, Shahrokhi M, Moradian R, Dovesi R (2015) On how differently the quasi-harmonic approximation works for two isostructural crystals: thermal properties of periclase and lime. J Chem Phys 142:044114

Eriksson G, Pelton AD (1993) Critical evaluation and optimization of the thermodynamic properties and phase diagrams of the CaO-$Al_2O_3$, $Al_2O_3$-$SiO_2$ and CaO-$Al_2O_3$-$SiO_2$ systems. Metall Trans B 24:807–816

Eriksson G, Wu P, Blander M, Pelton AD (1994) Critical evaluation and optimization of the thermodynamic properties and phase diagrams of the MnO-$SiO_2$ and CaO-$SiO_2$ systems. Canad Metall Trans 33:13–21

Essene E (1974) High-pressure transformations in $CaSiO_3$. Contrib Miner Petrol 45:247–250

Fawcett JJ, Yoder HS Jr (1966) Phase relations in the chlorites in the system MgO-$Al_2O_3$-$SiO_2$-$H_2O$. Am Miner 51:353–380

Fei Y, Bertka CM (1999) Phase transitions in the Earth's mantle and mantle mineralogy. In: Fei Y, Bertka CM, Mysen BO (eds) Mantle petrology: field observations and high pressure experimentation. Special Publication No 6, Geochemical Society, Houston, pp 189–207

Fei Y, Saxena SK, Navrotsky A (1990) Internally consistent thermodynamic data and equilibrium phase relations for compounds in the system MgO-$SiO_2$ at high pressure and high temperature. J Geophys Res 95:6915–6928

Fei Y, Mao HK, Shu J, Parthasarathy G, Bassett WA (1992) Simultaneous high-P, high-T X ray diffraction study of β-$(Mg,Fe)_2SiO_4$ to 26 GPa and 900 K. J Geophys Res 97:4489–4495

Fenner CL (1913) The stability relations of the silica minerals. Am J Sci 36:331–384

Finger LM, Hazen RM, Prewitt CT (1991) Crystal structures of $Mg_{12}Si_4O_{19}(OH)_2$ (phase B) and $Mg_{14}Si_5O_{24}$ (phase AnhB). Am Miner 76:1–7

Flood H, Grjotheim K (1952) Thermodynamic calculation of slag equilibria. J Inorg Steel Inst 171:64–71

Foster WR (1950) Synthetic sapphirine and its stability relations in the system $MgO-Al_2O_3-SiO_2$. J Geol 58:135–151
Forest J (1971) Connaissance de l'Orthosilicate de Calcium (Knowledge of Dicalcium Silicate). SFMC Bull 94:118–137
Fowler, RH, Guggenheim, EA (1939) Statistical thermodynamics. Cambridge University Press, 248p
Fortes AD (2019) Thermal expansion of the $Al_2SiO_5$ polymorphs, kyanite, andalusite and sillimanite, between 10 and 1573 K determined using time-offlight neutron powder diffraction. Phys Chem Min 46:687–704
Fredrickson GH, Liu AJ, Bates FS (1994) Entropic corrections to the Flory-Huggins theory of polymer blends: architectural and conformational effects. Macromol 27:2503–2511
Friedrich A, Kunz M, Winkler B, Le Bihan T (2004) High-pressure behavior of sillimanite and kyanite: compressibility, decomposition and indications of a new high-pressure phase Z. Kristallogr 219:324–329
Gale A, Dalton CA, Langmuir CH, Su Y, Schilling JG (2013) The mean composition of ocean ridge basalts. Geochem Geophys Geosys 14:489–758
Galkin VM, Dorochev AM, Babitch YuV (1987) Termitcheskoe raschirenie koesita. Izvestia AN SSSR. Geochimia 1645–1646
Ganguly J (2008) Thermodynamics in earth and planetary sciences. Springer ISBN 978-3-540-77305-4, 501p
Ganguly J, Kennedy GC (1974) The energetics of natural garnet solid solutions I mixing of the aluminosilicate end members. Contrib Miner Petrol 48:137–148
Ganguly J, Ghose S (1979) Aluminous orthopyroxene: order-disorder, thermodynamic properties, and petrological implications. Contrib Miner Petrol 69:375–385
Ganguly J, Saxena SK (1984) Mixing properties of aluminosilicate garnets: constraints from natural and experimental data and applications to geothermo-barometry. Am Miner 69:78–87
Ganguly J, Saxena SK (1987) Mixtures and mineral reactions. Springer-Verlag, Berlin-Heidelgerg-New York, 291p
Ganguly J, Cheng W (1994) Thermodynamics of (Ca,Mg,Fe,Mn)-garnet solid solution: new experiments, optimized data set, and applications to thermo-barometry. In: IMA, 16th general meeting (Abstracts), Pisa, Italy
Ganguly J, Frost DJ (2006) Stability of anhydrous phase B: Experimental studies and implications for phase relations in subducting slab and the X discontinuity in the mantle. J Geophys Res 111:B06203
Ganguly J, Cheng W, O'Neill HStC (1993) Syntheses, volume and structural changes of garnets in the pyrope-grossular join: implications for stability and mixing properties. Am Miner 78:583–593
Gasparik T (1984) Experimental study of subsolidus phase relations and mineral properties of pyroxene in the system $CaO-Al_2O_3-SiO_2$. Geochim Cosmochim Acta 48:2537–2545
Gasparik T (1992) Melting experiments on the enstatite-pyrope join at 80–152 kilobar. J Geophys Res Solid Earth 97:15181–15188
Gasparik T (2003) Phase diagrams for geoscientists an atlas of the Earth's interior. Springer, Berlin-Heidelberg-New York, p 462p
Gasparik T, Newton RC ((1984) The reversed alumina contents o orthopyroxene in equilibrium with spinel and forsterite in the system $MgO-Al_2O_3-SiO_2$. Contrib Miner Petrol 85:186–196
Gasparik T, Wolf K, Smith CM (1994) Experimental determination of phase relations in the $CaSiO_3$ system from 8 to 15 GPa. Am Miner 79:1219–1222
Gatta D, Nestola F, Walter JM (2006) On the thermo-elastic behaviour of kyanite: a neutron powder diffraction study up to 1200 °C. Min Mag 70:309–317
Gatti C, Ottonello G, Richet P (2012) Energetics and bonding in aluminosilicate ring with alkali metal and alkaline-earth metal charge-compensating cations. J Phys Chem A 116:8584–8598

Gelsdorf G, Muller-Heisse H, Schwiete HE (1958) Einlangerungsveruche an synthetischem Mullit und substitutionsveruche mit Galliumoxyd und Germaniumoxyd teil II. Arch Eisenhüttenw 29:513–519
Geisinger KL, Spackman MA, Gibbs GV (1987) Exploration of structure, electron density distribution, and bonding in coesite with Fourier and pseudo atom refinement methods using single-crystal X-ray diffraction data. J Phys Chem 91:3237–3244
Ghiorso MS, Carmichael ISE, Moret LK (1979) Inverted high-temperature quartz. Unit cell parameters and properties of the alfa-beta inversion. Contrib Miner Petrol 68:307–323
Glasser FP, Osborn EF (1958) Phase equilibrium studies in the system $CaO-Cr_2O_3-SiO_2$. J Am Ceram Soc 41:358–367
Glorieux B, Millot F, Rifflet J-C, Coutures J-P (1999) Density of superheated and undercooled liquid alumina by a contactless method. Int J Thermophys 20:1085–1094
Göbbels M, Woermann E, Jung J (1995) The Al-rich part of the system $CaO-Al_2O_3-MgO$ Part I phase relationships. J Solid State Chem 120:358–363
Goldsmith JR (1980) The melting and breakdown reactions of anorthite at high pressures and temperatures. Am Miner 65:272–284
Goodwin DW, Lindop AJ (1970) The crystal structure of $CaO{\cdot}2Al_2O_3$. Acta Cryst B 26:1230–1235
Grattan-Bellew PE (1978) Quartz-tridymite transition under hydrothermal conditions. Exp Miner 11:128–139
Greig JW (1927) Immiscibility in silicate melts Part I. Am J Sci 13:1–44
Guggenheim EA (1935) The statistical mechanics of regular solutions. Proc R Soc Lond Ser A 148:304–312
Gutt W. (1968) Manufacture of portland cement from phosphate raw materials. In: Building research station paper CP90/68
Haas JL, Robinson GR Jr, Hemingway BS (1981) Thermodynamic tabulations for selected phases in the system $CaO-Al_2O_3-SiO_2-H_2O$ at 101.325 kPa (1 atm) between 273.15 and 1800 K. USGS Bull, Series Number 1544
Hageman VBM, Oonk HAJ (1986) Liquid immiscibility in the $SiO_2$ + MgO, $SiO_2$ + SrO, $SiO_2$ + $LaO_3$, and $SiO_2$ + $Y_2O_3$ systems. Phys Chem Glasses 27:194–198
Hageman VBM, van den Berg GJK, Janssen HJ, Oonk HAJ (1986) A reinvestigation of liquid immiscibility in the $SiO_2$-CaO system. Phys Chem Glasses 27:100–106
Haiber M, Ballone P, Parrinello M (1997) Structure and dynamics of protonated $Mg_2SiO_4$: Ab-initio molecular dynamics study. Am Miner 82:913–922
Hamano K, Sato T, Nakagawa Z (1986) Properties of mullite powder prepared by coprecipitation and microstructure of fired bodies. Yogyo Kyokaishi 94:818–822
Hanemann H, Schrader A (1943) Beitrag zur Untersuchung von Viertstofflegierungen. Z Metallk 35:185–193
Hariya Y, Keddedy GC (1968) Equiibrium study of anorthite under high pressure and high temperature. Am J Sci 266:193–203
Haselton HT, Newton RC (1980) Thermodynamics of pyrope-grossular garnets and their stabilities at high temperatures and pressures. J Geophys Res 85:6973–6982
Haussühl E, Vinograd V, Krenzel TF, Schreuer J, Wilson DJ, Jan Ottinger J (2011) High temperature elastic properties of Mg-cordierite: experimental studies and atomistic simulations. Z Kristallogr 226:236–253
Harker RI, Tuttle OF (1956) The lower limit of stability of akermanite ($Ca_3MgSi_2O_7$). Am J Sci 254:468–478
Hartfield T, Richmond C, Ford WF, White J (1970) Compatibility relationships between the periclase and silicate phases in magnesite refractories at high temperatures. Trans Br Ceram Soc 69:53–58
Hays JF (1966) Lime-alumina-silica. Carn Inst Wash Yb 65:234–239
Hazen R (1976) Effect of temperature and pressure on the crystal structure of forsterite. Am Miner 61:1280–1293
Hazen RM, Downs RT, Finger LW, Ko J (1993) Crystal chemistry of ferromagensian silicate spinels Evidence of Mg-Si disorder. Am Miner 78:1320–1323

Hazen RM, Weinberger MB, Yang H, Prewitt CT (2000) Comparative high-pressure crystal chemistry of wadsleyite, β-(Mg1-xFex)$SiO_4$, with x=0 and 025. Am Miner 85:770–777
Helffrich G, Connolly JAD (2009) Physical contradictions and remedies using simple polythermal equations of state. Am Miner 94:1616–1619
Helgeson HC, Delany J, Bird DK (1978) Summary and critique of the thermodynamic properties of rock-forming minerals. Am J Sci 278A:1–229
Hemingway BS, Bohlen SR, Hankins WB, Westrum EF Jr, Kuskov OL (1998), Heat capacity and thermodynamic properties for coesite and jadeite, reexamination of the quartz-coesite equilibrium boundary. Am Miner 83:409–418
Henney JW, Jones JWS (1969) Solid solubility of calcia and silica in magnesia and its effect on the magnesia-calcia-silica system at 1750 °C. Trans Br Ceram Soc 68:210–213
Henriksen AF, Kingery WD (1979) The solid solubility of $Sc_2O_3$ $Al_2O_3$ $Cr_2O_3$ $SiO_2$ and $ZrO_2$ in MgO. Ceram Int 5:11–17
Hensen BJ, Schmid R, Wood BJ (1975) Activity-composition relationship for pyrope-grossular garnet. Contrib Miner Petrol 54:161-166
Hernández ER, Brodholt J, Alfè D (2015) Structural, vibrational and thermodynamic properties of $Mg_2SiO_4$ and $MgSiO_3$ minerals from first-principles simulations. Phys Chem Miner 240:1–24
Herzberg CT (1983) The reaction forsterite + cordierite = aluminous orthopyroxene + spinel in the system MgO-$Al_2O_3$-$SiO_2$. Contrib Miner Petrol 84:84–90
Herzberg C, Zhang J (1996) Melting experiments on anhydrous peridotite KLB-1: compositions of magmas in the upper mantle and transition zone. J Geophys Res Solid Earth 101:8271–8295
Herzberg CH, Gasparik T (1989) Melting experiments on chondrite at high pressures: stability of anhydrous B. Eos Trans Am Geophys Union 70:484
Higgins JB, Ribbe PH (1979) Sapphirine II: a neutron and X-ray diffraction study of $(Mg\text{-}Al)^{VI}$ and $(Si\text{-}Al)^{IV}$ ordering in monoclinic sapphirine. Contrib Miner Petrol 68:357–368
Hillert M, Jansson B, Sundman B (1988) Application of the compound energy model to oxide systems. Z Metallkd 79:81–87
Hillert M, Sundman B, Wang X (1989) A thermodynamic evaluation of the $Al_2O_3$-$SiO_2$ system. TRITA-MAC-402, KTH, Stockholm
Hillert M, Sundman B, Wang X (1990) An assessment of the CaO-$SiO_2$ system. Metall Trans B 21:303–312
Hillert M, Sundman B, Wang X, Barry T (1991) A reevaluation of the rankinite phase in the CaO-$SiO_2$ system. Calphad 15:53–58
Hofmeister AM (2004) Physical properties of calcium aluminates from vibrational spectroscopy. Geochim Cosmochim Acta 68:4721–4726
Holdaway MJ (1971) Stabilityof andalusite and the aluminosilicate phase diagram. Am J Sci 271:97–131
Holland KG, Ahrens TJ (1997) Melting of (Mg,Fe)2Sio4 at the core-mantle boundary of the Earth. Science 275:1623–1625
Holland TJB, Powell R (1998) An internally consistent thermodynamic data set for phases of petrological interest. J Metamorp Geol 16:309–343
Holland TJB, Navrotsky A, Newton RC (1979) Thermodynamic parameters of $CaMgSi_2O_6$, $Mg_2Si_2O_6$ pyroxenes based on regular solution and cooperative disordering models. Contrib Miner Petrol 69:337–344
Holm JL, Kleppa OL, Westrum EF (1967) Thermodynamics of polymorphic transitions in silica Thermal properties from 5 to 1070 °K and pressure-temperature stability fields for coesite and stishovite. Geochim Cosmochim Acta 31:2289–2307
Horiuchi H, Hirano M, Ito E, Matsui Y (1982) $MgSiO_3$ (ilmenite-type): Single crystal X-ray diffraction study. Am Miner 67:788–793
Horiuchi H, Ito E, Weidner DJ (1987) Perovskite-type $MgSiO_3$: single-crystal X-ray diffraction study. Am Miner 72:357–360
Hosieni KR, Howald RA, Scanlon MW (1985) Thermodynamics of the lambda transition and the equation of state of quartz. Am Miner 70:782–793

Howald RA, Eliezer I (1978) Thermodynamic properties of mullite. J Phys Chem 82:2199–2204
Hu X, Liu X, He Q, Wang H, Qin S, Ren L, Wu CM, Chang L (2018) Thermal expansion of andalusite and sillimanite ay ambient pressure: a powder X-ray diffraction study up to 1000 °C. Min Mag 7:363–374
Huang W-L, Wyllie PJ (1975) Melting and subsolidus melting relationships for $CaSiO_3$ to 35 kilobars pressure. Am Miner 60:213–217
Huang W, Hillert M, Wang X (1995) Thermodynamic assessment of the CaO-MgO-$SiO_2$ system. Metall Mater Trans A 26:2293–2310
Huckenholz HG, Yoder HS Jr (1974) The gehlenite-$H_2O$ and wollastonite-$H_2O$ systems. Carnegie Inst Wash Yb 73:440–443
Huckenholz HG, Holzl E, Lindhuber W (1975) Grossularite, its solidus and liquidus relations in the CaO-$Al_2O_3$-$SiO_2$-$H_2O$ system up to 10 kbar. Newes Yb Miner Abh 124:1–46
Hudon P, Jung I-H, Baker DR (2002) Melting of β-quartz up to 20 GPa and thermodynamic optimization of the silica liquidus up to 60 GPa. Phys Earth Planet Inter 130:159–174
Hudon P, Jung I-H, Baker DR (2004) Effect of pressure on liquid-liquid miscibilità gaps: a case study of the systems CaO-$SiO_2$, MgO-$SiO_2$, and $CaMgSi_2O_6$-$SiO_2$. J Geophys Res B 109:03207
Hudon P, Jung I-H, Baker DR (2005) Experimental investigation and optimization of thermodynamic properties and phase diagrams in the systems CaO-$SiO_2$, MgO-$SiO_2$, $CaMgSi_2O_6$-$SiO_2$ $CaMgSi_2O_6$-$Mg_2SiO_4$ to 10 GPa. J Petrol 46:1859–1880
Hugh-Jones D (1997) Thermal expansion of $MgSiO_3$ and $FeSiO_3$ ortho- and clinopyroxenes. Am Miner 82:689–696
Ikeda K, Suginohara Y, Yanagase T (1967) Effects of oxide addictions for the density and surface tension of barium silicate melts. Nippon Nogei Kagakukaishi 31:547–552
Inoue T, Irifune T, Higo Y, Sanchira T, Sueda Y, Yamada A, Shinmei T, Yamazaki D, Ando J, Funakoshi K, Utsumi W (2006) The phase boundary between wadsleyite and ringwoodite in $Mg_2SiO_4$ determined by in situ X-ray diffraction. Phys Chem Miner 33:106–114
Isaak DG, Anderson OL, Goto T (1989) Elasticity of single-crystal forsterite measured to 1700 K. J Geophys Res 94:5895–5906
Ito H, Kawada K, Akimoto S (1974) Thermah expansion of stishovite. Phys Earth Planet Inter 8:277–281
Iyi N, Göbbels M, Matsui Y (1995) The Al-rich part of the system CaO-$Al_2O_3$-MgO Part II structure refinement of two new magnesioplumbite-related phases. J Solid State Chem 120:364–371
Kambayashi S, Kato E (1984) A thermodynamic study of (magnesium oxide + silicon dioxide) by mass spectrometry at 1973 K. J Chem Thermodyn 16:241–248
Kantor PB, Lazareva LS, Kandiba VV, Formichov EM (1962) Experimental determination of the enthalpy, temperature, and heat of phase transition of the a-modification of $Al_2O_3$ in the temperature range 1200° to 2500° K. Ukr Fiz Zh 7:205–211
Kanzaki M (1990) Melting of silica up to 7 GPa. J Am Ceram Soc 73:3706–3707
Kanzaki M, Stebbins JF, Xue X (1991) Characterization of quenched high pressure phases in $CaSiO_3$ system by XRD and $^{29}$Si NMR. Geophys Res Lett 18:463–466
Karki BB, Stixrude L, Clark SJ, Warren MC, Ackland GJ, Crain J (1997) Structure and elasticity of MgO at high pressure. Am Miner 82:51–60
Karki BB, Wentzcovitch RM, de Gironcoli S, Baroni S (2000) High-pressure lattice dynamics and thermoelasticity of MgO. Phys Rev B 61:8793–8800
Karki BB, Bhattarai D, Stixrude L (2006) First principles calculations of the structural, dynamical and electronic properties of liquid MgO. Phys Rev B 73:174208-1–174217
Karki BB, Bhattarai B, Stixrude L (2007) First-principles simulations of liquid silica: structural and dynamical behaviour at high pressure. Phys Rev B 76:104205
Kato K, Saalfeld H (1968) Verfeinerung der kristallstruktur von $CaO(Al_2O_3)_6$ Neues Jahrb. Miner Abh 109:192–200
Katsura T, Ito E (1989) The system $Mg_2SiO_4$-$Fe_2SiO_4$ at high pressures and temperatures: precise determination of stabilities of olivine, modified spinel, and spinel. J Geophys Res 94:15663–15670

Keith ML, Schairer JF (1952) The stability field of sapphirine in the system $MgO$-$Al_2O_3$-$SiO_2$. J Geol 60:181–186

Kelley KK (1962) Heats and Free energies of formation of anhydrous silicates. Bur Mines Rep Invest 5901:4–7

Kennedy GC, Wasserburg GJ, Heard HC, Newton RC (1962) The upper three phase region in the system $SiO_2$-H2O. Am J Sci 260:501–521

Kerrich DM, Ghent ED (1979) P-T-$X_{CO2}$ relations of equilibria in the system $CaO$-$Al_2O_3$-$SiO_2$-$H_2O$ In: Zharikov VA, Fonarev VI, Korikovshii VA (eds) Problems in physico-chemical petrology. Acad Sci USSR 2:32–52

Kerrick DM, Henninger SG (1984) The andalusite sillimanite equilibrium revisited. Geol Soc Am Abstr Prog 16:558

Kiefer B, Stixrude L, Wentzcovitch R (1999) Normal and inverse ringwoodite at high pressures. Am Miner 84:288–293

Kiefer B, Stixrude L, Hafner J, Kresse G (2001) Structure and elasticity of wadsleyite at high pressures. Am Miner 86:1387–1395

Kieffer SW (1979) Thermodynamics and lattice vibrations of minerals: 1 mineral heat capacities and their relationships to simple lattice vibrational models. Rev Geophys Space Phys 17:1–19

Kieffer SW (1979) Thermodynamics and lattice vibrations of minerals: 2 vibrational characteristics of silicates. Rev Geophys Space Phys 17:20–34

King EG (1955) Heat capacities at low temperatures and at 29815K of crystalline calcium and magnesium aluminates. J Phys Chem 59:218–219

Kirshenbaum AD, Cahill JA (1960) The density of liquid aluminium oxide. J Inorg Nucl Chem 14:283–287

Kirfe A, Will G, Arndt J (1979) A new phase of coesite $SiO_2$. Z Krist 149:315–326

Kiseleva IA (1976) Thermodynamic parameters of the natural ordered and the synthetic disordered sapphirines. Geokhimiya 2:189–201

Klug FJ, Prochazka S, Doremus RH (1987) Alumina-silica phase diagram in the mullite region. J Am Ceram Soc 70:750–759

Kojitani H, Nishimura K, Kubo A, Sakashita M, Aoki K, Akaogi M (2003) Raman spectroscopy and heat capacity measurement of calcium ferrite type $MgAl_2O_4$ and $CaAl_2O_4$. Phys Chem Miner 30:409–415

Kojitani H, Terata S, Ohsawa M, Mori D, Inaguma Y, Akaogi M (2017) Experimental and thermodynamic investigations on the stability of $Mg_{14}Si5O_{24}$ anhydrous phase B with relevance to $Mg_2SiO_4$ forsterite, wadsleyite, and ringwoodite. Am Miner 102:2032–2044

Konopicky K (1956) Equilibrium diagram of the system $SiO_2$-$Al_2O_3$. Bull Soc Fr Céram 33:3–6

Kosmatka S (2012) Cement. Kirk-Othmer encyclopedia of chemical technology, pp 1–40. ISBN 978-0-471-48494-3

Kracek FC (1930) The cristobalite liquidus in the alkali oxide-silica systems and the heat of fusion of cristobalite. J Am Chem Soc 52:1436–1442

Kröger FA (1964) The chemistry of imperfect crystals. North Holland Pub Co, Amsterdam, 1039p

Krupka KM, Kerrick DM, Ito J (1985) Low-temperature heat capacities and derived thermodynamic properties of anthophyllite diopside enstatite bronzite and wollastonite. Am Miner 70:249–260

Krupka KM, Kerrich DM, Ito J (1985) High-temperature heat capacities and derived thermodynamic properties of anthophyllite, diopside, dolomite, enstatite, and bronzite. Am Miner 70:261–271

Krupka KM, Hemingway BS, Robie RA, Kerrick DM (1985) High-temperature heat capacities and derived thermodynamic properties of anthophyllite diopside dolomite enstatite bronzite talc tremolite and wollastonite. Am Miner 70:261–271

Kubaschewski O, Evans EL, Alcock CB (1967) Metallurgical thermochemistry. Pergamon, London, p 495p

Kudo R, Ito E (1996) Melting relations in the system $Mg_4Si_4O_{12}$ (En)-$Mg_3Al_2Si_3O_{12}$ (Py) at high pressures. Phys Earth Planet Inter 96:159–169

Kung J, Li B, Uchida T, Wang Y, Neuville D, Liebermann RC (2004) In situ measurements of sound velocities and densities across the orthopyroxene → high-pressure clinopyroxene transition in $MgSiO_3$ at high pressure. Phys Earth Planet Inter 147:27–44

Kung J, Li BS, Uchida T, Wang YB (2005) In-situ elasticity measurement for the unquenchable high-pressure clinopyroxene phase: implication for the upper mantle. Geophys Res Lett 32:L01307

Kuno H, Hess HH (1953) Unit cell dimensions of clinoenstatite and pigeonite in relation to other common clinopyroxenes. Am J Sci 251:741–752

Kuroda K, Irifune T, Inoue T, Nishiyama N, Miyashita M, Funakoshi K, Utsumi W (2000) Determination of phase boundary between ilmenite and perovskite in $MgSiO_3$ by in situ X-ray diffraction and quench experiments. Phys Chem Miner 27:523–532

Kushiro I (1964) Wollastonite-pseudowollastonite inversion. Carnegie Inst Wash YB 63:83–84

Kushiro I (1969) The system Forsterite–Diopside–Silica with and without water at high pressures. Am J Sci 267:269–294

Kushiro I (1972) Determination of liquidus relations in synthetic silicate systems with electron probe analysis: the system Forsterite–Diopside–Silica at 1 atmosphere. Am Miner 57:1260–1271

Jackson I (1976) Melting of the silica isotypes $SiO_2$, $BeF_2$ and $GeO_2$ at elevated pressures. Phys Earth Planet Inter 13:218–231

Jackson J, Sinogeikin SV, Bass JD (2000) Sound velocities and elastic properties of c-$Mg_2SiO_4$ to 873 K by Brillouin spectroscopy. Am Miner 85:296–330

Jacobs MHG, De Jong BHWS (2005) An investigation into thermodynamic consistency of data for the olivine, wadsleyite and ringwoodite form of (Mg,Fe)2Sio4. Geochim Cosmochim Acta 69:4361–4375

Jahn S, Martoňak R (2009) Phase behavior of protoenstatite at high pressure studied by atomistic simulations. Am Miner 94:950–956

Jochym PT, Parlinski K, Krzywiec P (2004) Elastic tensor of the forsterite ($Mg_2SiO_4$) under pressure. Comput Mater Sci 29:414–418

Jorgensen JD (1978) Compression mechanisms in α-quartz structures—$SiO_2$ and $GeO_2$. J Appl Phys 49:5473–5478

Joswig W, Paulus EF, Winkler B, Milman V (2003) The crystal structure of $CaSiO_3$-walstromite, a special isomorph of wollastonite-II. Z Kristallogr 218:811–818

Jung I-H, Decterov SA, Pelton AD (2004) Critical thermodynamic evaluation and optimization of the MgO-$Al_2O_3$, CaO-MgO-$Al_2O_3$, and MgO-$Al_2O_3$-$SiO_2$ systems. JPED 25:329–345

Jung I-H, Decterov SA, Pelton AD (2005) Critical thermodynamic evaluation and optimization of the CaO-MgO-$SiO_2$ system. J Eur Ceram Soc 25:313–333

Lacks DJ, Rear DB, Van Orman JA (2007) Molecular dynamics investigation of the viscosity, chemical diffusivities and partial molar volumes of liquids along the MgO-$SiO_2$ join as functions of pressure. Geochim Cosmochim Acta 71:1312–1323

Lagerqvist K, Wallmark S, Westgren A (1937) X-ray study of the systems CaO-$Al_2O_3$ and SrO-$Al_2O_3$. Z Anorg Allg Chem 234:1–16

Landau LD, Lifshitz EM (1980) Statistical physics Part 1, 3rd edn. Butterworth-Heinemann, Oxford, UK

Landron C, Hennet L, Jenkins TE, Greaves GN, Coutures JP, Soper AK (2001) Liquid alumina: detailed atomic coordination determined from neutron diffraction data using empirical potential structure refinement. Phys Rev Lett 86:4839–4842

Langstaff D, Gunn M, Greaves GN, Marsing A, Kargl F (2013) Aerodynamic levitator furnace for measuring thermophysical properties of refractory liquids. Rev Sci Instrum 84:124901

Lejus AM (1964) Sur la formation a haute temperature de spineless non stochiometriques et de phase derives. Rev Hautes Temp Et Réfract 1:53–95

Lejus AM, Collongues R (1962) Sur la formation a haute temperature de phases type alumine δ dans plusieurs systems à base d'alumine. Compt Rend Acad Sci 251:2780–2781

Levien L, Prewitt CT (1981) High-pressure crystal structure and compressibility of coesite. Am Miner 66:324–333

Li B, Liebermann RC (2007) Indoor seismology by probing the Earth's interior by using sound velocity measurements at high pressures and temperatures. Proc Natl Acd Sci 104:9145–9150
Li BS, Rigden SM, Liebermann RC (1996) Elasticity of stishovite at high pressure. Phys Earth Planet Inter 96:113–127
Li B, Liebermann RC, Weidner D (1998) Elastic moduli of wasdleyite ($\beta$-$Mg_2SiO_4$) to 7 gigapascals and 873 kelvin. Science 281:675–677
Li L, Weidner DJ, Wentzcovitch RM, Da Silva CRS (2007) Vibrational and thermodynamic properties of forsterite at mantle conditions. J Geophys Res B 112:05206-1–5208
Levien L, Prewitt CT (1981) High-pressure crystal structure and compressibility of coesite. Am Miner 66:324–333
Levien L, Prewitt CT, Weidner DL (1980) Structure and elastic properties of quartz at pressure. Am Miner 65:920–930
Liebske C, Frost DJ (2012) Melting phase relations in the MgO-$MgSiO_3$ system between 16 and 26 GPa: implications for melting in the Earth's deep interior. Earth Planet Sci Lett 345:159–170
Lindsley DH, Grover JE, Davidson PM (1981) The thermodynamics of the $Mg_2Si_2O_6$-$CaMgSi_2O_6$ join: a review and a new model. In: Saxena SK (ed) Advances in physical geochemistry, vol 1. Springer, New York, pp 149–175
Liu LG (1974) Disproportionation of kyanite to corundum plus stishovite at high pressure and temperature. Earth Planet Sci Lett 24:224–228
Liu J, Zhang J, Flesh L, Li B, Weidner DJ, Liebermann D (1999) Thermal equation of state of stishovite. Phys Earth Planet in 112:257–266
Liu X, Shieh SR, Fleet ME, Zhang L (2009) Compressibility of a natural kyanite to 17.5 GPa. Progr Nat Sci 19:1281–1286
Longhi J, Boudreau AE (1980) The orthoenstatite liquidus field in the system Forsterite-Diopside-Silica at one atmosphere. Am Miner 65:563–573
Louisnathan S (1969) Refinement of the crystal structure of gehlenite. Can Miner 10:822–837
Majumdar AJ (1964) The quaternary phase in high-alumina-cement. Br Ceram Trans J 63:347–364
Manara D, Böhler R, Capriotti L, Quaini A, Bao Z, Boboridis K, Luzzi L, Janssen A, Pöml P, Eloirdi R, Konings RJM (2014) On the melting behavior of calcium monoxide under different atmospheres: a laser heating study. J Eur Ceram Soc 34:1623–1636
Manzano H, Dolado JS, Ayuela A (2009) Structural, mechanical, and reactivity properties of tricalcium aluminate using first-principles calculations. J Am Ceram Soc 92:897–902
Mao H, Sundman B, Wang Z, Saxena SK (2001) Volumetric properties and phase relations of silica—thermodynamic assessment. J Alloys Compd 327:253–262
Mao H, Fabrichnaya O, Selleby M, Sundman B (2005) Thermodynamic assessment of MgO-$Al_2O_3$-$SiO_2$ system. J Mater Res 20:975–986
Mao H, Selleby M, Sundman B (2005) Phase equilibria and thermodynamics in the $Al_2O_3$–$SiO_2$ system modeling of mullite and liquid. J Am Ceram Soc 88:2544–2551
Mao H, Hillert M, Selleby M, Sundman B (2006) Thermodynamic assessment of the CaO-$Al_2O_3$-$SiO_2$ system. J Am Ceram Soc 89:298–308
Matsui Y, Syono Y (1968) Unit cell dimensions of some synthetic olivine group solid solutions. Geochem. J 2:51–59
McCarthy MI, Harrison NM (1994) Ab initio determination of the bulk properties of MgO. Phys Rev B 49:8574–8582
McNally RN, Peters FI, Ribbe PH (1961) Laboratory furnace for studies in controlled atmospheres: melting points of MgO, in a $N_2$ atmosphere and of $Cr_2O_3$ in $N_2$ and in air atmospheres. J Am Ceram Soc 44:491–493
Meng Y, Weidner DJ, Gwanmesia DG, Liebermann RC, Vaughan MT, Wang Y, Leinenweber K, Pacalo RE, Yaganeh-Haeri A, Zhao Y (1993) In situ high P-T X Ray diffraction studies on three polymorphs ($\alpha$, $\beta$, $\gamma$) of $Mg_2SiO_4$. J Geophys Res 98:22199–22207
Meng Y, Fei Y, Weidner D, Gwanmesia GD, Hu J (1994) Hydrostatic compression of $\gamma$-$Mg_2SiO_4$ to mantle pressures and 700 K: thermal equation of state and related thermodynamic properties. Phys Chem Miner 21:407–412

Merlini M, Gemm M, Artioli G (2005) Thermal expansion and phase transitions in åkermanite and gehlenite. Phys Chem Miner 32:189–196

Merlini M, Gemmi M, Artioli G (2009) Thermal expansion and phase transitions in åkermanite and gehlenite. Phys Chem Miner 32:189–196

Merlini M, Gemmi M, Hanfland M, Crichton W (2009) High-pressure behavior of åkermanite and gehlenite and phase stability of the normal structure in melilites. Am Miner 94:704–709

Merlino S (1973) Polymorphism in sapphirine. Contrib Miner Petrol 41:23–29

Merlino S (1980) Crystal structure of sapphirine-1Tc. Z Kristallogr 151:91–100

Ming LC, Manghnani MH, Kim YH, Usha-Devi S, Xu JA, Ito E (1992) Thermal expansion studies of $(Mg,Fe)_2SiO_4$-spinels using synchrotron radiation In: Saxena SK (ed) Thermodynamic data, systematics and estimation. Advances in physical geochemistry, vol 10. Springer, New York, pp 315–334

Mirwald PW (1982) A High-pressure phase transition in cordierite. Am Miner 67:277–283

Mirwald PW, Massonne H-J (1980) The low-high quartz and quartz-coesite transition to 40 kbar between 600° and 1600°C and some reconnaissance data on the effect of $NaAlO_2$ component on the low quartz-coesite transition. J Geophys Res 85:6983–6990

Mondal P, Jeffery JW (1975) The crystal structure of tricalcium aluminate, $Ca_3Al_2O_6$. Acta Crystallogr Sect B 31:689–697

Montanari B, Civalleri B, Zicovich-Wilson CM, Dovesi R (2006) Influence of the exchange-correlation functional in all-electron calculations of the vibrational frequencies of corundum ($\alpha$-$Al_2O_3$). Int J Quant Chem 106:1703–1714

Moore PB (1969) The crystal structure of sapphirine. Am miner 54:31–49

Moore PB, Araki T (1972) Atomic arrangement of merwinite, $Ca_3Mg[SiO_4]_2$, an unusual dense-packed structure of geophysical interest. Am Miner 57:1355–1374

Moretti R, Ottonello G (2003) A polymeric approach to the sulfide capacity of silicate slags and melts. Metall Trans B 34:399–410

Morimoto N, Appleman DE, Evans HT Jr (1959) The crystal st6ructures of clinoenstatite and pigeonite. Z Kristallogr 114:120–147

Morse SA (1980) Basalts and phase diagrams. Springer, New York-Heidelberg-Berlin, 493p

Mosenfelder JL, Asimow PD, Ahrens TJ (2007) Thermodynamic properties of $Mg_2SiO_4$ liquid at ultra-high pressures from shock measurements to 200 GPa on forsterite and wadsleyite. J Geophys Res, B 112:JB004364

Mosesman MA, Pitzer KS (1941) Thermodynamic properties of the crystalline forms of silica. J Am Chem Soc 63:2348–2356

Murakami M, Hirose K, Kawamura K, Sata N, Ohishi Y (2004) Post-perovskite phase transition in $MgSiO_3$. Science 304:855–858

Nada R, Nicholas JB, McCarthy MI, Hess AC (1996) Basis sets for Ab initio periodic Hartree-Fock studies of zeolite/adsorbate interactions: He, Ne, and Ar in silica sodalite. Int J Quant Chem 60:809–820

Natali M, Attene M, Ottonello G (2011) Modeling liquidus hypersurfaces through simplicial complexes In: Puppo E, Brogni A, De Floriani E (eds) Eurographics Italian chapter conference the eurographics association

Navrotsky A, Wechsler BA, Geisinger K, Seifert F (1986) Thermochemistry of $MgAl_2O_4$-$Al_{8/3}O_4$ defect spinels. J Am Ceram Soc 69:418–422

Neuvonen KJ (1952) Thermocheminal investigation of the akermanite-gehlenite series. Fin Geol Comm Bull 158:1–50

Neuvonen KJ (1952) Heat of formation of merwinite and monticellite. Am J Sci, Bowen 2:373–380

Newton RC (1966a) Kyanite-Sillimanite equilibrium at 750 °C. Science 151:1222–1225

Newton RC (1966) Some calc-silicate equilibrium relations. Am J Sci 264:204–222

Newton RC (1972) An experimental determination of the high pressure stability limits of magnesian cordierite under wet and dry conditions. J Geol 80:398–420

Newton RC, Charlu TV, Kleppa OJ (1977) Thermochemistry of high pressure garnets and clinopyroxenes in the system $CaO$-$MgO$-$Al_2O_3$-$SiO_2$. Geochim Cosmochim Acta 41:369–377

Newton RC, Geiger CA, Kleppa OJ , Brousse C (1986) Thermochemistry of binary and ternary garnet solid solutions. IMA Abs Program, 186p

Nickel KG, Brey G (1984) Subsolidus orthpyroxene-clinopyroxene systematic in the sustem CaO-MgO-$SiO_2$, to 60 kbar: a re-evaluation of the regular solution model. Contrib Miner Petrol 87:35–42

Niesel K (1972) The importance of the $\alpha'_L$–$\alpha'_H$, transition in the polymorphism of Dicalcium Silicate. Silic Indus 37:136–138

Nityanand N, Fine HA (1983) The effect of TiO2 additions and oxygen potential on liquidus temperatures of some CaO-$Al_2O_3$ melts. Metall Trans B 14:685–692

Noel Y, Catti M, D'Arco P, Dovesi R (2006) The vibrational frequencies of forsterite $Mg_2SiO_4$: an all-electron ab-initio study with the CRYSTAL code. Phys Chem Miner 33:383–393

Nurse RW, Welch JH, Majumdar AJ (1965) The CaO-$Al_2O_3$ system in a moisture-free atmosphere. Trans Br Ceram Soc 64:323–332

Oganov AR, Dorogokupets PI (2003) All-electron and pseudopotential study of MgO: equation of state, anharmonicity, and stability. Phys Rev B 67:224110

Okada K, Otsuka N (1987) Change in chemical-composition of mullite formed from $2SiO_2$–$3Al_2O_3$ xerogel during the formation progress. J Am Ceram Soc 70:245–247

Okamura FPS, Ghose S, Ohashi H (1974) Structure and crystal chemistry of calcium Tschermak's pyroxene $CaAlAlSiO_6$. Am Miner 59:549–557

Ol'shanskii YI (1951) Equilibrium of two immiscible liquids in silicate systems of alkali earth metals. Dokl Akad Nauk 76:93–96

Olinger B (1976) The compression of stishovite. J Geophys Res 81:5241–5248

Olinger B, Halleck PM (1976) The compression of α-quartz. J Geophys Res 81:5711–5714

Ono A, Akaogi M, Kojitani H, Yamashita K, Kobayashi M (2009) High-pressure phase relations and thermodynamic properties of hexagonal aluminous phase and calcium-ferrite phase in the systems $NaAlSiO_4$-$MgAl_2O_4$ and $CaAl_2O_4$-$MgAl_2O_4$. Phys Earth Planet Inter 174:39–49

Osborn EF( 1942) The system $CaSiO_3$-diopside-anorthite. Am J Sci 240:751–788

Osborn EF (1943) The compound merwinite ($3CaO{\cdot}MgO{\cdot}2SiO_2$) and its stability relations within the system CaO-MgO-$SiO_2$ (preliminary report). J Am Ceram Soc 26:321–332

Osborn EF, Muan A (1964) Phase diagrams for ceramists. Levin EM, Robbins CR, McMurdie HF (eds) The American Ceramic Society, Columubus, OH, vol 1, 219p

Osborn EF, Schairer JF (1941) The ternary system pseudowollastonite-akermanite-gehlenite. Am J Sci 239:715–763

Ostrovsky IA (1966) PT-diagram of the system $SiO_2$-$H_2O$. Geol J 5:127–134

Ottonello G (1987) Energies and interactions in binary (Pbnm) orthosilicates: a born parameterization. Geochim Cosmochim Acta 51:3119–3135

Ottonello G (1992) Interactions and mixing properties in the (C2/c) clinopyroxene quadrilateral. Contrib Miner Petrol 111:53–60

Ottonello G (1997) Principles of geochemistry. Columbia University Press, New York-Chicester-West Sussex

Ottonello, G (2001) Thermodynamic constraints arising from the polymeric approach to silicate slags: the system CaO-FeO-$SiO_2$ as an example. J Non-Cryst Solids 282:72–85

Ottonello G (2005) Chemical interactions and configurational disorder in silicate melts. Ann Geophys 48:561–581

Ottonello G (2024a) Quantum geochemistry. Springer Nature, Heidelberg, 1123p

Ottonello G (2024) First principles thermodynamic assessment of the MgO-$SiO_2$ system. Rend Lincei Phys Sci Nat. https://doi.org/10.1007/s12210-024-01258-5

Ottonello G, Richet P (2014) The solvation radius of silicate melts based on the solubility of noble gases and scaled particle theory. J Chem Phys 140:044506-1–44509

Ottonello G, Della Giusta A, Dal Negro A, Baccarin F (1992) A structure- energy model for C2/ c pyroxenes in the system Na-Mg-Ca-Mn-Fe-Al-Cr-Ti-Si-O. In: Saxena SK (ed) Advances in physical geochemistry, vol 10. Springer, Berlin-Heidelberg-New York, pp 194–238

Ottonello G, Bokreta M, Sciuto PF (1996) Parameterization of energy and interactions in garnets: end-member properties. Am Miner 81:429–447

Ottonello G, Vetuschi Zuccolini M, Civalleri B (2009) Thermo-chemical and thermo-physical properties of stishovite: an ab-initio all-electron investigation. Calphad 33:457–468

Ottonello G, Civalleri B, Ganguly J, Vetuschi Zuccolini M, Noel Y (2009) Thermophysical properties of α–β–γ polymorphs of $Mg_2SiO_4$: a computational study. Phys Chem Miner 36:87–106

Ottonello G, Vetuschi Zuccolini M, Belmonte D (2010) The vibrational behavior of silica clusters at the glass transition: Ab-initio calculations and thermodynamic implications. J Chem Phys 133(104508):1–12

Ottonello G, Civalleri B, Ganguly J, Perger WF, Belmonte D, Vetuschi Zuccolini M (2010) Thermo-chemical and thermo-physical properties of the high-pressure phase anhydrous B ($Mg_{14}Si_5O_{24}$): an ab-initio all-electron investigation. Am Miner 95:563–573

Ottonello G, Attene M, Ameglio D, Belmonte D, Vetuschi Zuccolini M, Natali M (2013) Thermodynamic investigation of the $CaO$-$Al_2O_3$-$SiO_2$ system at high P and T through polymer chemistry and convex-hull techniques. Chem Geol 346:81–92

Ouchi Y, Yoshida T, Kato E (1977) Densities of ternary lead-silicate melts. Nippon Nogei Kagakukaishi 41:865–874

Pacalo REG, Gasparik T (1990) Reversals of the orthoenstatite-clinoenstatite transition at high pressures and high temperatures. J Geophys Res 95:90

Palko JW, Sayir A, Sinogeikin SV, Kriver WM, Bass JD (2002) Complete elastic tensor for mullite (~$25Al_2O_3{\cdot}SiO_2$) to high temperatures measured from textured fibers. J Am Ceram Soc 85:2005–2012

Palmer DC, Downs RT (1991) High pressure behaviour of cristobalite revealed by single crystal X-ray diffraction. Eos Trans AGU 72(44):486–638

Pankratz LB, Kelley KK (1964) High temperature heat contents of andalusite, kyanite, and sillimanite. US Bur Mines Rep Invent 6370:1–7

Paradis P-F, Ishikawa T, Saita Y, Yoda S (2004) Non-contact thermophysical property measurements of liquid and undercooled alumina. Jpn J Appl Phys 43:1496–1500

Parlinski K, Li ZQ, Kawazoe Y (1997) First-principles determination of the soft mode in cubic $ZrO_2$. Phys Rev Lett 78:4063–4066

Pascale F, Zicovich-Wilson CM, Orlando R, Roetti C, Ugliengo P, Dovesi R, Pascual-Ahuir JL, Silla E (2005) Vibration frequencies of $Mg_3Al_2Si_3O_{12}$ pyrope an Ab initio study with the CRYSTAL code. J Phys Chem 109:6146–6152

Pelton AD, Blander M (1986) Thermodynamic analysis of ordered liquid solutions by a modified quasichemical approach—application to silicate slags. Metall Trans B 17:805–815

Perdew JP, Wang Y (1986) Accurate and simple density functional for the electronic exchange energy: generalized gradient approximation. Phys Rev B 33:8800–8802

Perdew JP, Wang Y (1992) Accurate and simple analytic representation of the electron-gas correlation energy. Phys Rev B 45:13244–13249

Perdew JP, Zunger A (1981) Self-interaction correction to density-functional approximations for many-electron systems. Phys Rev B 23:5048–5079

Perkins D III, Holland TJB, Newton RC (1981) The $Al_2O_3$ contents of enstatite in equilibrium with garnet in the system $MgO$-$Al_2O_3$-$SiO_2$ at 15–40 kbar and 900°–1600°C. Contrib Miner Petrol 78:99–109

Piekarz P, Jochym PT, Parlinski K, Lazewski J (2002) High-pressure and thermal properties of γ-$Mg_2SiO_4$ from first-principles calculations. J Chem Phys 117:3340–3344

Pierotti RA (1976) A scaled particle theory of aqueous and nonaqueous solutions. Chem Rev 76:717–726

Prencipe M, Mantovani L, Tribaudino M, Bersani D, Lottici PP (2012) The Raman spectrum of diopside: a comparison between Ab initio calculated and experimentally measured frequencies. Eur J Miner 24:457–464

Preparata FP, Shamos MI (1985) Computational geometry: an introduction. Springer, New York, p 498p

Presnall DC, Gasparik T (1990) Melting of enstatite ($MgSiO_3$) from 10 to 165 GPa and the forsterite ($Mg_2SiO_4$)—majorite ($MgSiO_3$) eutectic at 165 GPa: implications for the origin of the mantle. J Geophys Res B 95:15771–15777

Presnall DC, Dixon JR, O'Donnell TH, Dixon SA (1979) Generation of mid-ocean ridge tholeiites. J Petrol 20:1–55

Presnall DC, Weng YH, Milholland CS, Walter MJ (1998) Liquidus phase relations in the system MgO-$MgSiO_3$ at pressures up to 25 GPa—constraints on crystallization of a molten Hadean mantle. Phys Earth Planet Int 107:83–95

Ralph RL, Finger LW, Hazen RM, Ghose S (1984) Compressibility and crystal structure of andalusite at high pressure. Am Miner 69:513–519

Rankin GA, Merwin HE (1916) The ternary system CaO-$Al_2O_3$-MgO. J Am Ceram Soc 38:568–588

Rankin GA, Merwin HE (1918) The ternary system MgO-$Al_2O_3$-$SiO_2$. Am J Sci 45:301–325

Rankin GA, Wright FE (1915) The ternary system CaO-$Al_2O_3$-$SiO_2$. Am J Sci 39:1–79

Raz U (1983) Thermal and volumetric measurements on quartz and other substances at pressures up 1108 to 6 kbars and temperatures up to 700 °C. PhD thesis (unpublished), ETH Zürich

Redlich O, Kister AT (1948) Algebraic representation of thermodynamic properties and the classification of solutions. Ind Eng Chem 40:345–349

Remy C, Guyot F, Madon M (1995) High pressure polymorphism of dicalcium silicate $Ca_2SiO_4$ a transmission electron microscopy study. Phys Chem Miner 22:419–427

Remy C, Andrault D, Madon M (1997) High-temperature, high-pressure X-ray investigation of dicalcium silicate. J Am Ceram Soc 80:851–860

Reiss H (1965) Scaled particle methods in the statistical thermodynamics of fluids. Adv Chem Phys 9:1–84

Richardson SW, Bell PM, Gilbert MC (1968) Kyanite-sillimanite equilibrium between 700 and 1500 K. Am J Sci 266:513–541

Richet P, Bottinga Y (1985) Heat capacity of aluminum-free liquid silicates. Geochim Cosmochim Acta 49:471–483

Richet P, Ottonello G (2014) The Earth as a multiscale quantum-mechanical system. CR Geosci 346:317–325

Richet P, Bottinga Y, Deniélou L, Petitet JP, Téqui C (1982) Thermodynamic properties of quartz, cristobalite and amorphous $SiO_2$: drop calorimetry measurements between 1000 and 1800 K and a review from 0 to 2000 K. Geochim Cosmochim Acta 46:2639–2658

Richet P, Robie RA, Hemingway BS (1991) Thermodynamic properties of wollastonite, pseudowollastonite and $CaSiO_3$ glass and liquid. Eur J Miner 3:475–484

Richet P, Leclerc F, Benoist L (1993) Melting of forsterite and spinel, with implications for the glass transition of $Mg_2SiO_4$ liquid. Geeophys Res Lett 20:1675–1678

Ridgen SM, Jackson I (1991) Elasticity of germanate and silicate spinels at high pressure. J Geophys Res 96:9999–10006

Rigden SM, Ahrens TJ, Stolper EM (1989) High-pressure equation of state of molten anorthite and diopside. J Geophys Res 94:9508–9522

Robie RA, Hemingway BS (1995) Thermodynamic properties of minerals and related substances at 29815 K and 1 Bar ($10^5$ Pascals) pressure and at higher temperatures. US Geol Surv Bull 2131, United States Government Printing office, Washington, 461p

Robie RA, Hemingway BS, Fisher JR (1978) Thermodynamic properties of minerals and related substances at 29815 K and 1 bar ($10^5$ pascal) pressure and at higher temperatures. United States Government Printing office, Washington, 456p

Rolin M, Pham HT (1965) Phase diagrams of mixtures not reacting with molybdenum [$Ca(AlO_2)^{2-}$, $NaAlO_2^-$ and $La_2O_3$-$Al_2O_3$]. Rev Int Hautes Témp Refract 2:175–185

Ross NL, Shu JF, Hazen RM, Gasparik T (1990) High-pressure crystal chemistry of stishovite. Am Miner 75:739–747

Roy DM, Roy R, Osborn EF (1953) The system $MgO-Al_2O_3-H_2O$ and influence of carbonate and nitrate ions on the phase equilibria. Am J Sci 251:337–361
Saalfeld H (1975) X-Ray investigation of single crystals of $\beta-Ca_2SiO_4$ (Larnite) at high temperatures. Am Miner 60:824–827i
Saafeld H, Jagodzinski H (1957) Segregation of Mg-Al spinels with an excess of $Al_2O_3$. Z Kristallogr 109:87–109
Salje E, Wruck B, Marais S (1991) Order parameter saturation at low temperatures—numerical results for displacive and O/D systems. Ferroelectrics 124:185–188
Saunderson DH, Peckham GE (1971) The lattice dynamics of calcium oxide. J Phys C 4:2009–2016
Saxena SK (1989) INSP and THERMO computer routines and corresponding data-base Uppsala University. Unpublished, Sweden
Saxena SK, Chatterjee N, Fei Y, Shen G (1993) Thermodynamic data on oxides and silicates an assessed data set based on thermochemistry and high pressure phase equilibrium. Springer, Berlin-Heidelberg-New York, p 164p
Schlaudt CM, Roy DM (1965) Crystalline solution in the system $MgO-Mg_2SiO_4-MgAl_2O_4$. J Am Ceram Soc 48:248–251
Schmidt MW, Poli S, Comodi P, Zanazzi PF (1997) High-pressure behavior of kyanite: decomposition of kyanite into stishovite and corundum. Am Miner 82:460–466
Schmitt D, Spehner J-C (2006) k-set polytopes and order-k Delaunay diagrams In: ISVD 2006, procs of the 3rd international symposium on voronoi diagrams in science and engineering, pp 173–185
Schreyer WJ, Schairer JF (1961) Compositions and structural states of anhydrous Mg-cordierites: a re-investigation of the central part of the system $MgO-Al_2O_3-SiO_2$. J Petro 2:324–406
Schreyer WJ, Seifert JF (1969) High-pressure phases in the system $MgO-A_{l2}O_3-SiO_2-H_2O$. Am J Sci 267A:407–443
Shmulovich KL (1974) Phase equilibria in the $CaO-Al_2O_3-SiO_2-CO_2$ system. Geochem Int 11:883–887
Schneider H, Eberhard E (1990) Thermal expansion of mullite. J Am Ceram Soc 73:2073–2076
Schneider H, Komarneni S (2005) Mullite. Wiley-VCH, Weinheim, p 487p
Schneider H, Schreuer J, Hildmann B (2008) Structure and properties of mullite—a review. J Eur Ceram Soc 28:329–344
Schoenitz M, Navrotsky A, Ross N (2001) Enthalpy of formation of $CaSi_2O_5$, a quenched high-pressure phase with pentacoordinate silicon. Phys Chem Miner 28:57–60
Schairer JF, Bowen NL (1955) The system $K_2O-Al_2O_3-SiO_2$. Am J Sci 253:681–746
Seifert F (1973) Stability of the assemblage cordierite-corundum in the system $MgO-Al_2O_3-SiO_2-H_2O$. Contrib Miner Petrol 41:171–178
Seifert F (1974) Stability of saffirine: a tudy of the aluminous part of the system $MgO-Al_2O_3-SiO_2-H_2O$. J Geol 82:173–204
Shears EC, Archibald WA (1954) Alumino-silicate refractories. Iron Steel (London) 27:26–30
Shen G, Lazor P (1995) Measurement of melting temperatures of some minerals under lower mantle pressures. J Geophys Res 100:17699–17713
Sheperd ES, Rankin GA, Wright FE (1911) Preliminary report on the ternary system $CaO-Al_2O_3-SiO_2$. J Ind Eng Chem 3:211–227
Shi P, Saxena SK, Zhang Z, Sundman B (1994) Thermodynamics of the Ca-Mg-Fe-Al-Si-O pyroxenes: I theoretical model and assessment of the Ca-Mg-Si-O system. Calphad 18:47–69
Shirasuka K, Yamaguchi G (1974) Precise measurement of the crystal data of the solid solution range of the defective spinel magnesium oxide n (aluminum oxide). Yogyo Kyokaishi 82:34–37
Shpil'rain EE, Kagan DN, Barkhatov LS (1972) Thermodynamic properties of the condensed phase of alumina near the melting point. High Temp-High Press 4:605–609
Singh BP, Srivastava SK, Dinesh K (2004) Relation between thermal expansivity and bulk modulus for ionic solids at high temperatures. Physica B 349:401–407
Skinner BJ (1966) Thermal expansion In: Clark SP (ed) Handbook of physical constants, vol 97. Geological Society of America Memoir, pp 75–95

Skola V (1937) The decomposition of mullite. Keramische Rundschau und Kunst-Keramik 45:188–215

Smart RM, Glasser FP (1976) Phase relations of cordierite and sapphirine in the system MgO-$Al_2O_3$-$SiO_2$. J Mater Sci 11:1459–1464

Smyth JR, lto J (1977) The synthesis and crystal structure of a magnesium-lithium-scandium protopyroxene. Am Miner 62:1252–1257

Smyth JR, Smith JV, Artioli G, Kvick A (1987) Crystal structures of coesite at 15 K and 298 K from neutron and X-ray single crystal data: test of bonding models. J Phys Chem 91:988–992

Spencer PJ (1973) The thermodynamic properties of silicates. Nat Phys Lab Div Chem Stand Rep N 21, Teddington, UK, 43p

Srivastava SK (2006) Analysis of thermoelastic constants of solids and minerals at high temperatures. J Phys Chem Solids 67:2275–2281

Stacey FD (1977) A thermal model of the Earth. Phys Earth Planet Int 15:341–348

Stangarone C (2016) Ab initio calculations of Raman and IR spectra of orthoenstatite and forsterite: lattice dynamics and modelling for planetary remote sensing. Dissertation, University of Parma wwwrepositoryuniprit@bitstream

Stangarone C, Tribaudino M, Prencipe M, Lottici PP (2016) Raman modes in Pbca enstatite ($Mg_2Si_2O_6$): an assignment by quantum mechanical calculation to interpret experimental results. J Raman Spectrosc 47:1247–1258

Staronka A, Pham H, Rolin M (1968) Étude du système silice-alumine par la méthode des courbes de refroidissement. Rev Int Hautes Témp Refract 5:111–115

Stebbins JF, Carmichael ISE, Moret LK (1984) Heat capacities and entropies of silicate liquids and glasses. Contrib Miner Petrol 86:131–148

Stixrude L, Karki BB (2005) Structure and freezing of $MgSiO_3$ liquid in Earth's lower mantle. Science 310:297–299

Stixrude L, de Koker N, Sun N, Mookherjee M, Karki BB (2009) Thermo-dynamics of silicate liquids in the deep Earth. Earth Planet Sci Lett 278:226–232

Strachan A, Çağin T, Goddard WA III (2001) Reply to "comment on 'phase diagram of MgO from density-functional theory and molecular-dynamics simulations.' Phys Rev B 63:096102

Sugiyama M, Endo S, Koto K (1987) The crystal structure of stishovite under pressure up to 6 GPa. Miner J 13:455–466

Suito K (1977) Phase relations of pure $Mg_2SiO_4$ up to 200 kilobars. In: Mangnahni MH, Akimoto S (eds) High-pressure research; applications in geophysics. Acacemic Press, SanDiego, California, pp 255–266

Sun XW, Song T, Chu YD, Liu ZJ, Zhang ZR, Chen QF (2010) The high-pressure melting curve of CaO. Solid State Comm 150:1785–1788

Suzuki I (1975) Thermal expansion of periclase and olivine, and their anharmonic properties. J Phys Earth 23:145–159

Suzuki I, Anderson OL (1983) Elasticity and thermal expansion of a natural garnet up to 1,000 K. J Phys Earth 31:3125–3138

Suzuki Y, Ohji T (2004) Anisotropic thermal expansion of calcium dialuminate ($CaAl_4O_7$) simulated by molecular dynamics. Ceram Int 30:57–61

Suzuki I, Ohtani E, Kumazawa M (1979) Thermal expansion of $\gamma$-$Mg_2SiO_4$. J Phys Earth 27:53–61

Suzuki I, Ohtani E, Kumazawa M (1980) Thermal expansion of modified spinel $\beta$-$Mg_2SiO_4$. J Phys Earth 28:273–280

Swamy V, Dubrovinsky LS (1997) Thermodynamic data for the phases in the $CaSiO_3$ system. Geochim Cosmochim Acta 61:1181–1191

Swainson IP, Dove MT (1995) On the thermal expansion of b-cristobalite. Phys Chem Miner 22:61–65

Swainson IP, Dove MT, Schmahl WW, Putnis A (1992) Neutron powder diffraction study of the akermanite-gehlenite solid solution series. Phys Chem Miner 19:185–195

Taylor D (1984) Thermal expansion. IV. Binary oxides with the silica structures. Br Ceram Trans J 83:129–134

Taylor JR, Dinsdale AT (1990) Thermodynamic and phase diagram data for the $CaO-SiO_2$ system. Calphad 14:71–88

Teter DM, Gibbs GV, Boisen MB Jr, Allan DC, Teter MP (1995) First-principles study of several hypothetical silica framework structures. Phys Rev B 52:8064–8073

Tewhey JD, Hess PC (1979) The two phase region in the $CaO-SiO_2$ system: experimental data and thermodynamic analysis. Phys Chem Glasses 20:41–53

Thiéblot L, Téqui C, Richet P (1999) High-temperature heat capacity of grossular ($Ca_3Al_2Si_3O_{12}$) enstatite ($MgSiO_3$) and titanite ($CaTiSiO_5$). Am Miner 84:848–855

Thompson AB, Wennemer M (1979) Heat capacities and inversions in tridymite, cristobalite and tridymite-cristobalite mixed phases. Am Miner 64:1018–1026

Thompson AB, Aerts M, Hack AC (2007) Liquid immiscibility in silicate melts and related systems. In: Liebscher A, Heinrich CA (eds) Fluid-Fluid Interactions; Rev Miner Geochem 65:99–127

Todd SS (1951) Low-temperature heat capacities and entropies at 29816 K of crystalline calcium orthosilicate, zinc orthosilicate and tricalcium silicate. J Am Chem Soc 73:3277–3278

Tomasi J, Persico M (1994) Molecular interactions in solution: an overview of methods based on continuous distributions of the solvent. Chem Rev 94:2027–2094

Tomlinson JW, Heynes MSR, Bokris JO'M (1958) The structure of liquid silicates Part 2—molar volumes and expansivities. Trans Faraday Soc 54:1822–1833

Toop GW, Samis CS (1962) Some new ionic concepts of silicate slags. Can Met Quart 1:129–152

Toop GW, Samis CS (1962) Activities of ions in silicate melts. Trans Met Soc AIME 224:878–887

Traverse JP, Foex M (1969) Investigation of the $ZrO_2$-CaO, and $ZrO_2$-SrO systems. High Temp-High Press 1:409–427

Troemel G, Fix W (1969) Subsolidus Relations in the System $2CaO{\cdot}SiO_2$-$3CaO{\cdot}P_2O_5$. Keram Rund 93:1–8

Trønnes RG, Frost DJ (2002) Peridotite melting and mineral-melt partitioning of major and minor elements at 22–245 GPa. Earth Planet Sci Lett 197:117–131

Tsuchiya T, Kawamura K (2001) Systematics of elasticity: Ab initio study in B1-type alkaline earth oxides. J Chem Phys 114:10086–10093

Ulmer P, Stalder R (2001) The Mg(Fe)SiO3 orthoenstatite-clinoenstatite transitions at high pressures and temperatures determined by Raman-spectroscopy on quenched samples. Am Miner 86:1267–1274

Vaidya SN, Bailey S, Pasternack T, Kennedy GC (1973) Compressibility of fifteen minerals to 45 kilobars. J Geophys Res Solid Earth 78:6893–6898

Viechnicki D, Schmid F, McCauley JW (1974) Liquidus-solidus determinations in the system $MgAl_2O_4$-$Al_2O_3$. J Am Ceram Soc 57:47–48

Viertel HM, Seifert E (1980) Thermal stability of defect spinels in the system $MgAl_2O_4$-$Al_2O_3$. Neues Jahrb Miner Abh 140:89–101

Vočadlo L, Price GD (1996) The melting of MgO-omputer calculations via molecular dynamics. Phys Chem Miner 23:42–49

Vogel R (1937) The heterogeneous equilibria. Handbook der Metallphysik, vol II. G. Masing, Leipzig, pp 669–698

Voskov AL, Voronin GF (2010) A universal method for calculating isobaric-isothermal sections of ternary system phase diagrams. Russ J Phys Chem 84:525–533

Wagman DD, Evans WH, Parker VB, Schumm RH, Halow I, Bailey SM, Churney KL, Nuttall RL (1982) The NBS tables of chemical thermodynamic properties, National Bureau of Standards (NBS). Washington, DC. J Phys Chem Ref Data 11(Suppl):21

Wagner VH (1932) Zur Thermochemie der Metalsilikate des Calciums und Magnesiums und des Diopsids. Z Anorg Allg Chem 208:1–22

Walter LS (1963) Experimental studies on Bowen's decarbonation series: I: P-T univariant equilibria of the "monticellite" and "akermanite" reactions. Am J Sci 261:488–500

Wang Y, Weidner DJ (1994) Thermoelasticity of $CaSiO_3$ perovskite and implications for the lower mantle. Geophys Res Lett 21:895–898

Wang Y, Weidner DJ, Guyot F (1996) Thermal equation of state of $CaSiO_3$ perovskite. J Geophys Res 101:661–672
Warner RD, Luth WC (1973) Two-phase data for the join Monticellite ($CaMgSiO_4$)—Forsterite ($Mg_2SiO_4$): experimental results and numerical analysis. Am Miner 58:998–1008
Watanabe H (1982) Thermochemical properties of synthetic high-pressure compounds relevant to the Earth's mantle. In: Akimoto H, Manghnani MH (eds) High pressure research in geophysics. Center for Academc Publications, Tokio, pp 441–464
Weidner DJ, Sawamoto H, Sasaki S, Kumazawa M (1984) Single-crystal elastic properties of the spinel phase of $Mg_2SiO_4$. J Geophys Res 89:7852–7860
Welch JH (1961) Ternary compound formation in the system CaO-$Al_2O_3$-MgO. Nature 4788:559–560
Welch JH, Gutt W (1959) Tricalcium silicate and its stability within the system CaO-$SiO_2$. J Am Ceram Soc 42:11–15
Weng Y, Presnall DC (2001) The system diopside—forsterite—enstatite at 51 GPa: a thernary model for melting in the mantle. Can Miner 39:299–308
Wentzcovitch RM (1991) Invariant molecular-dynamics approach to structural phase transitions. Phys Rev B 44:2358–2361
Wentzcovitch RM, Martins JL (1991) First principles molecular dynamics of Li: test of a new algorithm. Solid State Commun 78:831–834
Wentzcovitch RM, Wu Z, Carrier P (2010) First principles quasiharmonic thermoelasticity of mantle minerals In: Wentzcovitch RM and Stixrude L (eds) Theoretical and computational methods in mineral physics; Geophys Appl Rev Miner Geochem 71:99–128
White WP (1919) Silicate specific heats. Am J Sci 47:1–21
Whitney WP, Stubican VS (1971) Interdiffusion in the system MgO–$MgAl_2O_3$. J Am Ceram Soc 54:349–352
Windom KE, Boettcher AL (1976) The effect of reduced activity of anorthiteon the reaction grossular + quartz = anorthite + wollastonite: a model for plagioclase in the earth's lower crust and upper mantle. Am Miner 61:889–896
Wisnyi LG (1955) High alumina phases in the system lime-alumina. PhD thesis, Rutgers University, New Brunswick, New Jersey
Wood BJ (1978) Reactions involving anorthite and $CaAlSi_2O_6$ pyroxene at high pressures and temperatures. Am J Sci 278:930–942
Wood BJ (1988) Activity measurements and excess entropy-volume relationships for pyrope-grossular garnets. J Geol 96:721–729
Wu Z, Wentzcovitch RM (2007) Vibrational and thermodynamic properties of wadsleyite: a density functional study. J Geophys Res 112:B12202-1–11
Wu P, Eriksson G, Pelton AD (1993) Critical evaluation and optimization of the thermodynamic properties and phase diagrams of the CaO-FeO, CaO-MgO,CaO-MnO, FeO-MgO, FeO-MnO, and MgO-MnO systems. J Am Ceram Soc 76:2065–2075
Wu Z, Wentzcovitch RM, Umemoto K, Li B, Hirose K, Zheng J-C (2008) Pressure-volume-temperature relations in MgO: an ultrahigh pressure-temperature scale for planetary sciences applications. J Geophys Res 113:B06204-1-12
Xiong Z, Liu X, Shieh SR, Wang S, Chang L, Tang J, Hong X, Zhang Z, Wang H (2016) Some thermodynamic properties of larnite (b-Ca2SiO4) constrained by high T/P experiment and/or theoretical simulation. Am Miner 101:277–288
Yamada T, Yoshimura M, Somiya S (1986) Reinvestigation of the solidification point of CaO by digital pyrometry. J Am Ceram Soc 69:C243–C245
Yang N, Ghose S (1994) In-situ Fe-Mg order-disorder studies and thermodynamic properties of orthopyroxene (Mg,Fe)$2S_i2_O6$. Am Miner 79:633–643
Yang H, Downs RT, Finger LW, Hazen RM, Prewitt CT (1997) Compressibility and crystal structure of kyanite, $Al_2SiO_5$, at high pressure. Am Miner 82:467–474
Yang H, Hazen RM, Finger LW, Prewitt CT, Downs RT (1997) Compressibility and crystal structure of sillimanite, $Al_2SiO_5$, at high pressure. Phys Chem Miner 25:39–47

Yao J, Frost DJ, Steinle-Neumann G (2021) Lower mantle melting: experiments and thermodynamic modeling in the system $MgO\text{-}SiO_2$. J Geophys Res Solid Earth 126:1–11
Yoder HS Jr (1968) Akermanite and related melilite-bearing assemblages. Carnegie Inst Wash Yb 471:483
Yu YG, Wentzcovitch RM (2006) Density functional study of vibrational and thermodynamic properties of ringwoodite. J Geophys Res 111:B12202-1-8
Yu YG, Wentzcovitch RM, Angel RJ (2010) First principles study of thermodynamics and phase transition in low-pressure (P21/c) and high-pressure (C2/c) clinoenstatite $MgSiO_3$. J Geophys Res 115:B02201-1-10
Yu YG, Angel RJ, Ross NL, Gibbs GV (2013) Pressure impact on the structure, elasticity, and electron density distribution of $CaSi_2O_5$. Phys Rev B 87:184112
Yuan L, Ohtani E, Shibazaki Y, Ozawa S, Jin Z, Suzuki A, Frost DJ (2018) The stability of anhydrous phase B, $Mg_{14}Si_5O_{24}$, at mantle transition zone conditions. Phys Chem Miner 45:523–531
Zerr A, Boehler R (1993) Melting of (Mg,Fe)$Si_O$3-perovskite to 625 Kilobars: indication of a high melting temperature in the lower mantle. Science 262:553–555
Zerr A, Boehler R (1994) Constraints on the melting temperature of the lower mantle from high-pressure experiments on MgO and magnesiowüstite. Nature 371:506–508
Zhang L, Fei Y (2008) "Melting behavior of (Mg, Fe) O solid solutions at high pressure. Geophys Res Lett 35. https://doi.org/10.1029/2008GL034585
Zhang J, Liebermann RC, Gasparik T, Herzberg CT (1993) Melting and subsolidus relations of $SiO_2$ at 9-14 GPa. J Geophys Res 98:19,785-19,793
Zienert T, Fabrichnaya O (2013) Thermodynamic assessment and experiments in the system MgO–$Al_2O_3$. Calphad 40:1–9